A Guide to

Weather

A GUIDE TO

WEATHER

WILLIAM J. BURROUGHS, BOB CROWDER,
TED ROBERTSON, ELEANOR VALLIER-TALBOT,
RICHARD WHITAKER

CONSULTANT EDITOR
RICHARD WHITAKER

FOG CITY PRESS

Published by Fog City Press
814 Montgomery Street
San Francisco, CA 94133 USA

This edition 2005

Chief Executive Officer John Owen
President Terry Newell
Publisher Lynn Humphries
Managing Editor Janine Flew
Design Manager Helen Perks
Editorial Coordinator Jennifer Losco
Editorial Assistant Irene Mickaiel
Production Manager Caroline Webber
Production Coordinator Monique Layt
Business Manager Emily Jahn
Vice President International Sales Stuart Laurence

Project Editors Jenni Bruce, Scott Forbes
Assistant Editor Greg Hassall
Copy Editors Julia Cain, Gillian Hewitt, Dawn Titmus
Project Art Director Hilda Mendham
Designers Clive Collins, Clare Forte, Lena Lowe
Design Assistant Stephanie Cannon
Picture Research Gillian Manning

A catalog record for this book is available from
the Library of Congress, Washington, DC.

ISBN 1 740893 33 6

Color reproduction by Colourscan Co Pte Ltd
Printed by Kyodo Printing Co. (S'pore) Pte Ltd
Printed in Singapore

A Weldon Owen Production

This grand show is eternal. It is always sunrise somewhere; the dew is never all dried at once; a shower is forever falling; vapor is ever rising. Eternal sunrise, eternal sunset, eternal dawn and gloaming, on sea and continents and islands, each in its turn, as the round earth rolls.

JOHN MUIR (1838–1914),
Scottish-born American naturalist and writer

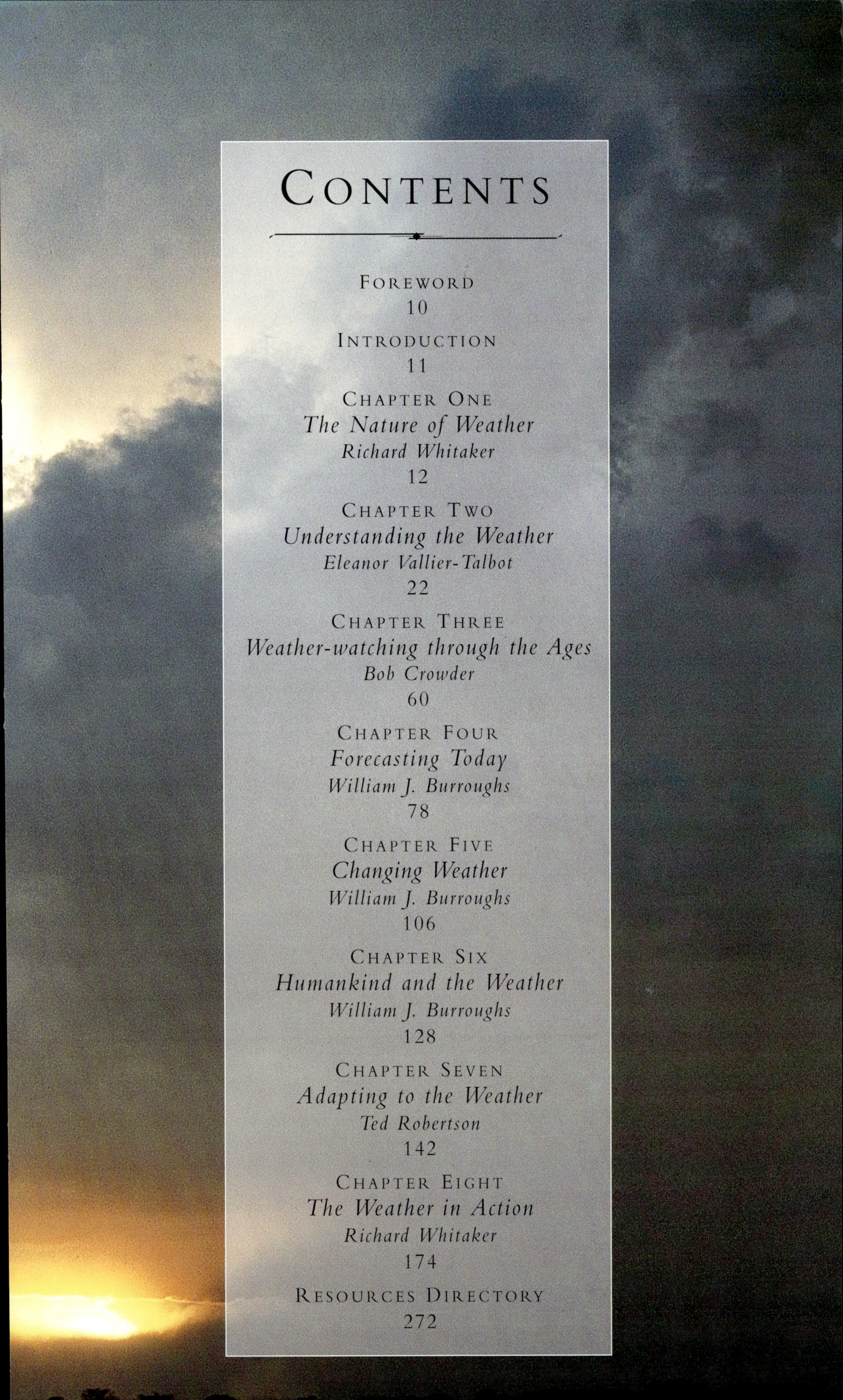

Contents

FOREWORD

The weather affects everyone. It is our constant companion—as tranquil, as turbulent, as wondrous and sometimes as unpredictable as life itself.

An appreciation of the beauty of nature has always been one of the privileges of humanity, but only this century has science succeeded in explaining many of the weather's mysteries. This book will help you to observe and understand the workings of weather and climate as revealed by the powerful tools of twentieth century meteorology.

Weather forecasting is one those rare activities that unite nations in a common endeavor from which people worldwide benefit daily. Through weather satellites and the combined efforts of the more than 180 member countries of the World Meteorological Organization, we can track the forces that control our weather and forecast their behavior up to a week or more ahead. Communities at risk can be warned of severe weather and virtually all sectors of society enabled to better plan their daily lives.

Meteorology is one of the oldest sciences and also one of the most challenging. And as the world begins to come to grips with global environmental issues ranging from acid rain to desertification, meteorology is assuming an important new role in underpinning the development of policies aimed at securing the long-term survival of the planet.

I hope that this book will enable you to experience the excitement of unraveling the mysteries of nature that is the privilege of all those who work with the weather.

JOHN W. ZILLMAN
Director, Australian Bureau of Meteorology

INTRODUCTION

The symmetry of a rainbow, the beauty of a cloud formation, the awesome power of a thunderstorm: We admire the aesthetic qualities of these everyday events, but how many of us wonder at the atmospheric conditions—the weather—that give rise to them? Yet the weather pervades all aspects of our lives: what we wear, our leisure pursuits, our moods. It even colors our idioms; we speak of being snowed under, or say that every cloud has a silver lining, or that we'll take a rain check.

A Guide to Weather invites us to take a closer look at this phenomenon that we take for granted. Mark Twain said, "The weather is always doing something," and that's true no matter where you go, for weather is a universal occurrence. But what it does, and when, is what defines our sense of time and place; and in that respect weather is local.

This book examines weather from both persectives. It discusses global patterns and influences, and details how to measure their effects on your immediate surroundings. Also featured is a field guide that profiles various aspects of the weather—such as tornadoes, clouds, lightning, dust devils, floods, auroras, and mirages—and explains when, why, and where they occur.

This book is part of a series that focuses on popular science and natural history. Beautiful to look at and easy to use, each volume draws on the knowledge and experience of experts in the field and presents a wealth of information in a lively and accessible way. Perhaps the best way to enjoy this book is to take it along on a picnic. Spread a blanket on the grass, lie on your back, and then name the clouds. It's another way of looking closely at this wonderful Earth.

THE EDITORS

CHAPTER ONE
THE NATURE *of* WEATHER

Joys come from simple and natural things, mists over meadows, sunlight on leaves, the path of the moon over water. Even rain and wind and stormy clouds bring joy …

Open Horizons,
SIGURD F. OLSON (1899–1982), American author and naturalist

A World *of* Weather

The way we live and all the myriad forms of life on our planet result from adaptations to a wide variety of weather patterns.

The weather may well be humankind's most widely discussed topic. Its effects are all-pervasive, ranging from the trivial issue of whether we should take an umbrella to work, to the tragedies that unfold during extreme weather such as drought and flood.

The weather dictates the sort of life we lead, the way we build our homes, the way we dress. It even influences our leisure pursuits. It is not surprising that the best skiers come from the snowfields of Europe and North America, and the world's top surfers come from places where the weather is warm and the waves large, such as Hawaii and Australia.

In conjunction with geological forces, the weather has shaped the landforms around us, and the kaleidoscopic variety of life on Earth reflects nature's myriad solutions to the range of meteorological conditions that have occurred throughout history.

Consider the Amazonian swamp forest where rainfall is so frequent that the trees are adapted to living underwater much of the time, and the conifers of northern forests whose shape allows snow to slide off their branches. The emperor penguin of the Antarctic can survive several weeks of temperatures as low as -80° F (-62° C) and winds of over 100 miles per hour (160 kph) while standing in winter darkness. The kangaroo rat of the American deserts is able to survive long periods by obtaining all the water it needs from seeds and other food.

KACHINA DOLL *of the Hopi people in Arizona (above left), whose eyes represent rain clouds. Power lines in southern Bangladesh brought down by a tropical storm (above), and traffic halted by snow in North America (right).*

There are seven or eight categories of phenomena in the world worth talking about, and one of them is weather.

Pilgrim at Tinker Creek,
Annie Dillard (b. 1945),
American writer

Feeding Ourselves

The weather directly affects our ability to feed ourselves. Even in today's technological societies, food supplies are vulnerable to the effects of the weather. Short spells of bad weather can disrupt transport and harvesting, and prolonged severe weather can lead to shortages and high prices. In the developing world, where food is often in short supply, drought and flood can result in people starving.

The most pressing issues facing us worldwide are environmental degradation and the population explosion. It is therefore vital that our agricultural techniques become more efficient, both to avoid further soil and water degradation and to feed ever-growing numbers of people.

Weather forecasting can greatly assist in this respect. For instance, long-term forecasts of rainfall patterns allow farmers to plan crop planting and crop rotation for maximum benefit. Short-term forecasts can help farmers decide to withhold pesticide spraying until rain has passed, thus avoiding environmentally unfriendly reapplications. And frost warnings can give citrus growers time to implement frost-prevention measures that may avert serious crop losses.

FOOD SUPPLIES *are directly affected by weather. Two-thousand-year-old rice terraces in the Philippines (above) and wheat fields in Washington, USA (right).*

WEATHER-WATCHERS

People who are able to forecast the weather have always been valued. Early weather-watchers were priests or shamans whose duties included both predicting the weather and changing it to suit their community's needs. They aimed to do this by directing rites and sacrifices designed to persuade the weather gods to act appropriately. Rain dances have been common to many peoples including Native Americans, central African tribes, and Australian Aborigines. Similarly, people have used prayers in attempts to influence the weather.

Over the centuries, a body of weather folklore has developed: sayings and proverbs based on centuries of weather observations. Handed down from generation to generation, many are still with us, such as:

Red sky in the morning
Is a sailor's sure warning,
Sky red at night
Is the sailor's delight.

In the West, science began to influence weather-watching from the fifteenth century, but meteorology was not accepted as a science until this century. Today it is a complex practice using computer modeling and satellite technology.

RED SQUIRRELS *(left) store nuts in fall to serve as a winter food supply.*

THE AMATEUR

While forecasting is clearly the domain of the expert, learning to recognize the processes at work in the skies can enrich us all. In addition, understanding weather patterns can enhance outdoor activities such as sailing or birding, and may help us avoid life-threatening situations such as flash floods and severe thunderstorms.

Perhaps most important of all, by learning how our activities affect the weather and how these changes affect the natural world, we can develop a fuller appreciation of our surroundings and learn to live in greater harmony with our environment.

WHAT *is* WEATHER?

Created by the heat of the Sun, the weather is a system of cycles and forces within the atmosphere that envelops the Earth.

The Earth is surrounded by an envelope of air called the atmosphere. The atmosphere is so thin that if you are in an airplane at 30,000 feet (9,000 m), over three-quarters of its air molecules will be below you. While other planets in our solar system have atmospheres, ours is unique in that its mixture of gases contains water vapor and it experiences a range of temperatures that enables water to exist in gaseous, liquid, and solid forms.

DAY-TO-DAY WEATHER

The term weather is used to describe day-to-day variations in our atmosphere. These are recorded by meteorologists as measurements of temperature, humidity, cloud cover, wind, and precipitation, and experienced by us as cold, heat, wind, rain, and so on.

The source of all these changes is the Sun. As the Earth spins on its axis at an angle of about 23.5 degrees and orbits the Sun, it is heated in a highly irregular manner. Equatorial areas receive more intense radiation than those near the poles, and, because of different heat absorption characteristics, landmasses heat up more than do the oceans.

The atmosphere constantly works toward equilibrium, attempting to "smooth out" temperature irregularities by carrying warm air from the equator toward the poles and cold air toward the equator. However, as this happens, the air is redirected by the Earth's rotation, slowed down by friction with the land and sea, and held within the confines of the atmosphere by gravity.

Together, these cycles and forces create complex, ever-changing patterns. Vast waves of air and giant swirls of cloud up to several thousand miles wide circulate around our planet, in the way that currents and eddies occur in a river as it rushes over rocks and reaches

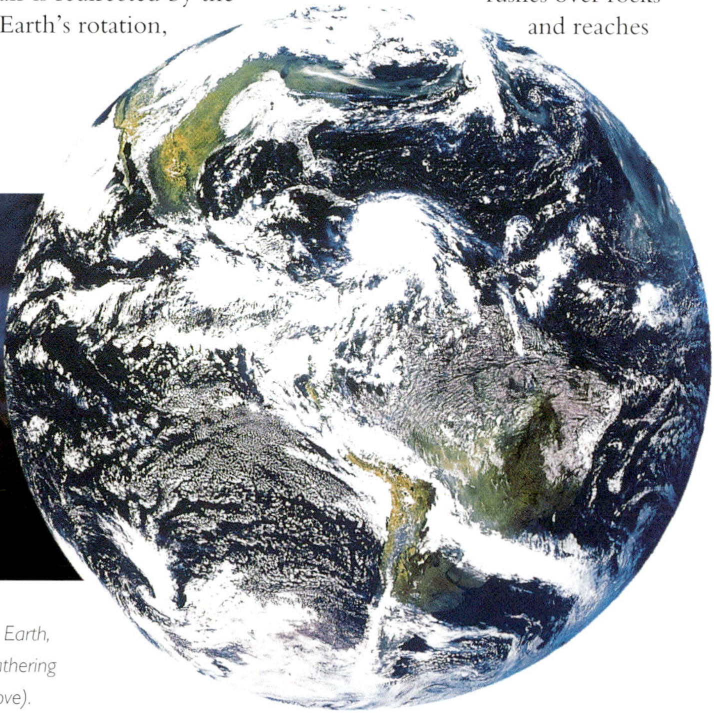

CLOUD MASSES *constantly swirl around the Earth, as this satellite image (right) clearly shows. Gathering storm clouds in North America's Midwest (above).*

SUNRISE IN SPACE *(above) showing the Earth's horizon and atmospheric effects caused by volcanic emissions. A tornado and its parent storm (left).*

deep pools. Only in the last century or so have we come to associate these swirls with areas of high and low pressure in the atmosphere that result in changes in the weather on the ground.

Average Weather

Climate is a long-term look at the weather, a synthesis of meteorological variables such as average monthly rainfall and maximum and minimum daily temperatures over a period of time. More than just averaging, such syntheses also take into account extremes and frequencies of occurrence. A minimum period of 30 years of recording is usually required to construct a climatic picture of an area. The longer the period, the more detailed the picture.

Other parameters are also investigated, including humidity, hours of sunshine, cloud cover, wind speed and direction, maximum wind gusts, and solar radiation. When combined, all this information produces a statement that we take to be the climate of the area.

Climate Change

Meteorologists are often asked: "Is our climate changing?" The answer is yes. There is plenty of evidence to show that the climate, like the weather, is constantly changing, although much more slowly.

Formal meteorological records vary in length. British Admiralty records, for instance, go back more than two centuries, and there are quite a number of countries that have records going back about 150 years. These can be augmented by historical material describing notable weather events. Such records, however, are of limited use for examining the climate over periods of thousands of years.

However, techniques are now available that enable us to reach much farther back in time than formal records permit. This is because nature has its own recording systems, such as tree rings, corals, and ice cores (see p. 114), and we are learning to read and interpret them with increasing skill. They show that our climate has changed throughout time, and numerous theories have been advanced to explain such changes.

Studies of weather statistics have also confirmed that human activities are modifying the atmosphere, particularly by releasing large quantities of "exotic" gases and particles into the air from motor vehicles and industrial processes. What effect will this have on climate? This question is now the subject of extensive international co-operation and research, and one of the main challenges of climatic studies is to separate the human effects from the natural.

ATMOSPHERIC PHENOMENA *include rainbows (right) and virga (above right)—raindrops that evaporate when they encounter a layer of warm air.*

WORLD CLIMATES

Numerous factors including wind regimes, sea-surface temperatures, and rainfall patterns must be taken into account when defining climatic zones.

Over the centuries, many methods of classifying global climatic zones have evolved. Most are based on the broad climatic zones bounded by the tropics of Cancer and Capricorn, and the Arctic and Antarctic circles.

The region between the two tropics is referred to as the low latitudes, and the climates that occur there are generally defined as tropical. The areas between the two tropics and the Arctic and Antarctic circles are known as the middle latitudes, and, for the most part, these have temperate climates. Between the Arctic and Antarctic circles and the poles lie the high latitudes, which have polar climates.

This broad classification is sometimes subdivided into maritime and continental zones, in recognition of the significant climatic differences between coastal and inland areas. While this distinction goes some way toward a more accurate reflection of actual conditions, it still does not account for the profound climatic effects of mountain ranges and ocean currents.

Just how significant these factors can be is made clear by comparing the climates of northern Newfoundland, in Canada, and northern Scotland. These areas are situated at similar latitudes and are both coastal. However, northern Newfoundland has a polar climate, while, as a result of the warming effect of the ocean current known as the Gulf Stream (see p. 38), Scotland has a temperate climate.

The climate classification presented here takes account of such factors, as well as the biomes, or vegetation zones, that occur around the world. As with all human classifications of naturally occurring phenomena, this system should be viewed as a general guide only. Within these regions climatic anomalies will exist, often as a result of local weather patterns.

COPING WITH THE CLIMATE

Seals (below) are among the few mammals that can tolerate a polar climate. Samburu in the semi-arid regions of Kenya, Africa (top right).

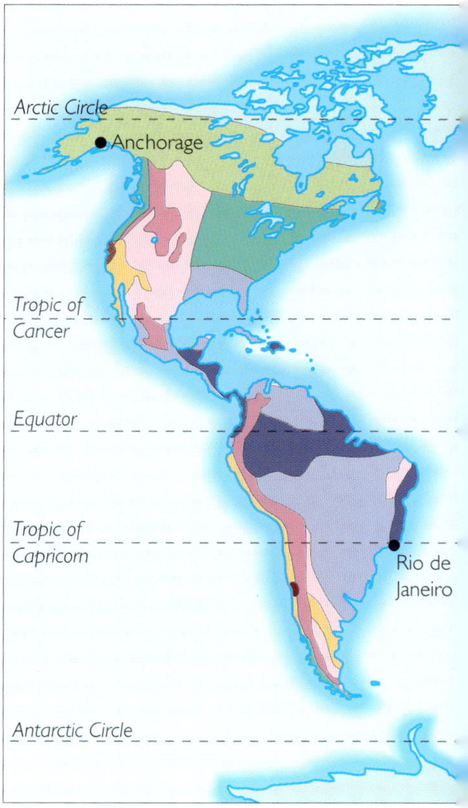

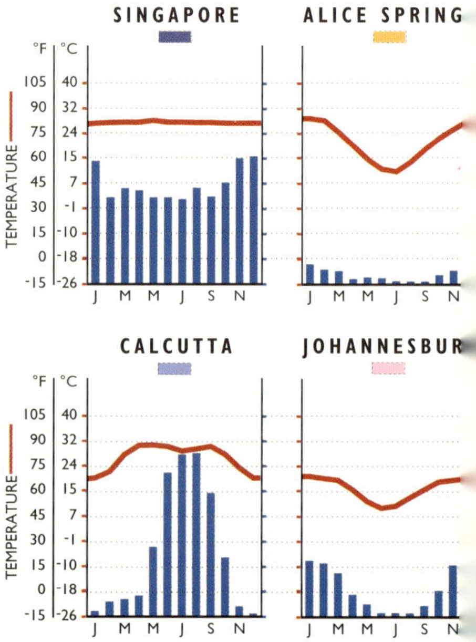

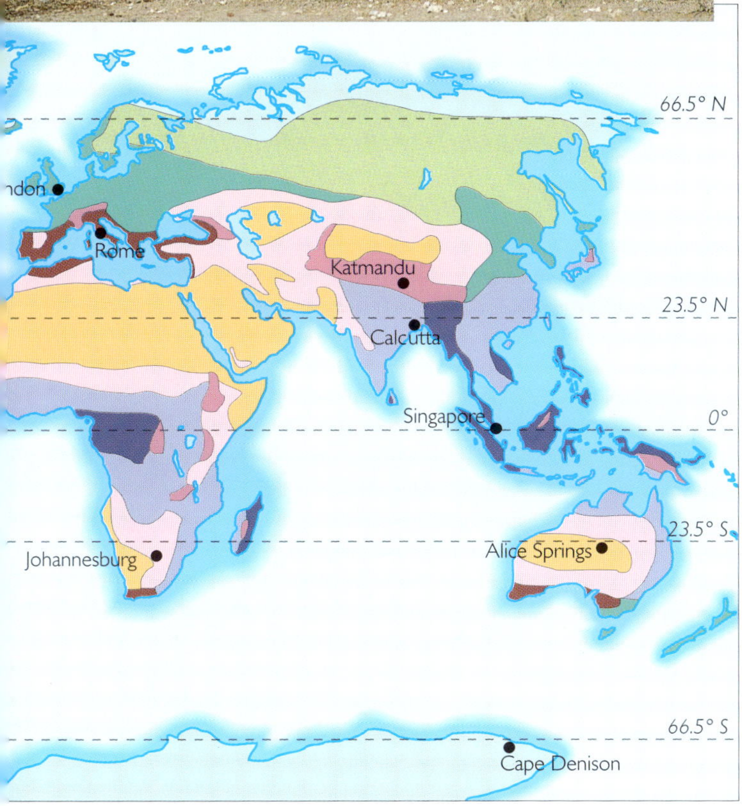

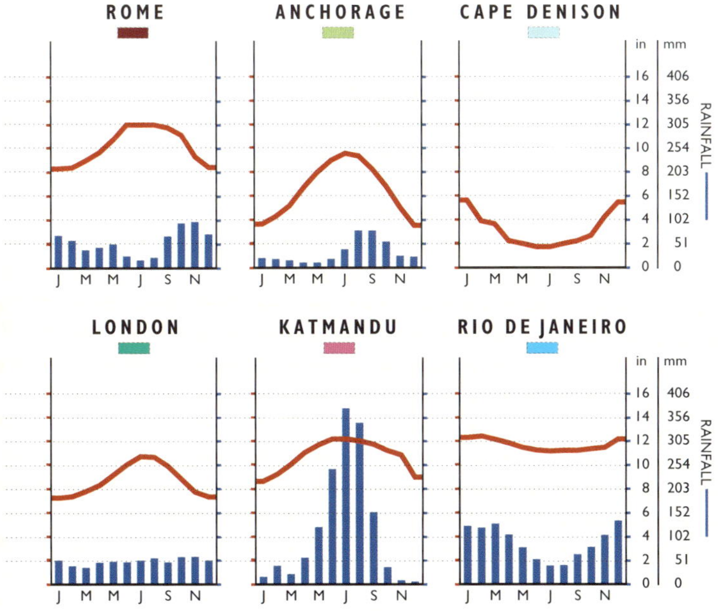

The Climatic Zones

Tropical High rainfall, consistently high temperatures and humidity, and a short dry season. See p. 146.

Subtropical Broader temperature range than tropical; wet and dry seasons of a similar length. See p. 150.

Arid Consistently low rainfall, and very large temperature fluctuations between night and day and between summer and winter months. See p. 152.

Semi-arid Conditions generally less extreme than arid regions. Higher rainfall and less marked temperature fluctuations between summer and winter months. See p. 156.

Mediterranean Hot, dry summers, and cool, wet winters. See p. 158.

Temperate Fairly uniform rainfall patterns, and four distinct seasons. Summers are warm and winters cold with snowfalls. See p. 160.

Northern temperate Similar characteristics to temperate, but winters are much longer, lasting up to nine months, and snowfall is much greater. See p. 162.

Mountain Temperatures much lower than low-level locations at similar latitudes. Regular snowfalls. Rainfall variable, depending on local rain-bearing winds. See p. 164.

Polar Extremely long and cold winter months, with slightly warmer summers. Frequent snow, but rainfall is negligible. See p. 168.

Coastal Narrower temperature range than nearby inland areas. Local weather dependent on sea-surface temperatures. See p. 172.

THE GRAPHS *on the left show average temperature (in red) and rainfall (in blue) for each location indicated on the map.*

PREDICTING *the* WEATHER

Increasingly efficient weather forecasts save money and lives. But will we ever be able to predict the weather accurately?

Many countries are currently pouring resources into weather forecasting, with the view that an efficient meteorological service produces benefits that are several times greater than the cost of the actual service.

Advanced weather services have been developed in the United States, the United Kingdom, many European countries, Japan, Australia, New Zealand, and South Africa. The People's Republic of China, with its heavy reliance on agriculture, has a weather service numbering about 65,000 people. Saudi Arabia is also placing a high priority on improving its meteorological services in the belief that it will create more efficient farming practices and increase the proportion of arable land in the country.

Because everything in the atmosphere is connected, successful weather prediction depends to a great extent on cooperation between countries. When forecasting for the United States, it is necessary to know the weather situation in Europe. Similarly, in Australia meteorologists need to have access to observations from South America.

This requirement to work together has been recognized by most countries and fostered by the World Meteorological Organization (WMO) (see p. 80). In this respect, the meteorological community has been one of the most successful of all international forums, maintaining contact and exchanging information since the early 1950s, despite a number of political crises. This has resulted in a vast global network of information and sophisticated technology being available to meteorologists for their daily operations.

Access to these resources is undoubtedly improving our weather-forecasting abilities, as are a number of other factors, such as our growing knowledge of the physical processes involved,

MEASUREMENTS *An instrument shelter (top). Launching an ozone-testing balloon in Antarctica (above). Mount Washington weather station, New Hampshire, USA (right).*

GLOBAL COOPERATION *is essential for successful forecasting. Meteorologists such as this one in Bracknell, England (right), contribute to the tracking of storms in other parts of the world. Destruction caused by Hurricane Andrew in Florida in 1992 (far right).*

more frequent observations by increasingly sophisticated meteorological satellites, and steady improvement in mathematical simulation of the weather. This last endeavor requires the processing of thousands of observations from weather stations around the world every few hours, as well as intermittent satellite information, and reports from aircraft and ships at sea. Modern supercomputers are required to handle this massive quantity of data.

Types of Forecast

These resources allow meteorologists to produce remarkably detailed weather forecasts. The precise amount of information that is included depends on the "range" of the forecast.

Short-range forecasts are made for the following 24 to 48 hours and usually include detailed predictions of temperature, cloudiness, wind speed and direction, and rainfall. Medium-range forecasts are made 48 hours to 10 days ahead and are more general, but may still include fairly detailed temperature and rainfall estimates. Long-range predictions, for up to three months ahead, are usually limited to whether rainfall or temperature is likely to be above or below average.

WEATHER SATELLITES *provide visual images, sense ground temperatures, and measure wind speeds and atmospheric moisture. This satellite view shows the Indian subcontinent and Sri Lanka.*

Everybody talks about the weather, but nobody does anything about it.

Mark Twain (1835–1910), American writer and humorist

The Limits of Forecasting

Recently, it has become clear that weather is only ever predictable up to a point. This is because the weather passes from modes of high predictability into phases of uncertainty where even short-range forecasts become a challenge.

When the weather exhibits extreme sensitivity to initial conditions—the classic signature of chaos—it flips into an unpredictable phase. It is as if the atmosphere teeters on a razor's edge, and it is theoretically possible for a butterfly to flap its wings in Brazil and, by the tiny movement of air produced, trigger a series of events that culminates in a tornado over Ohio or a typhoon off the coast of Japan.

Getting It Right

Given these limits, how much can we expect forecasts to improve? In most state-of-the-art weather services around the world, accuracies in prediction of about 85 percent are routinely achieved in forecasting 24 hours ahead. Although long-range forecasts are somewhat less dependable, significant accuracy is also being demonstrated in forecasts of up to five or six days ahead, and as improved mathematical simulations of the weather become available, accurate forecasts for up to 10 days ahead may become routine in the near future.

I wield the flail of the lashing hail,
And whiten the green plains under,
And then again I dissolve it in rain,
And laugh as I pass in thunder.

The Cloud,
PERCY BYSSHE SHELLEY (1792–1822), English poet

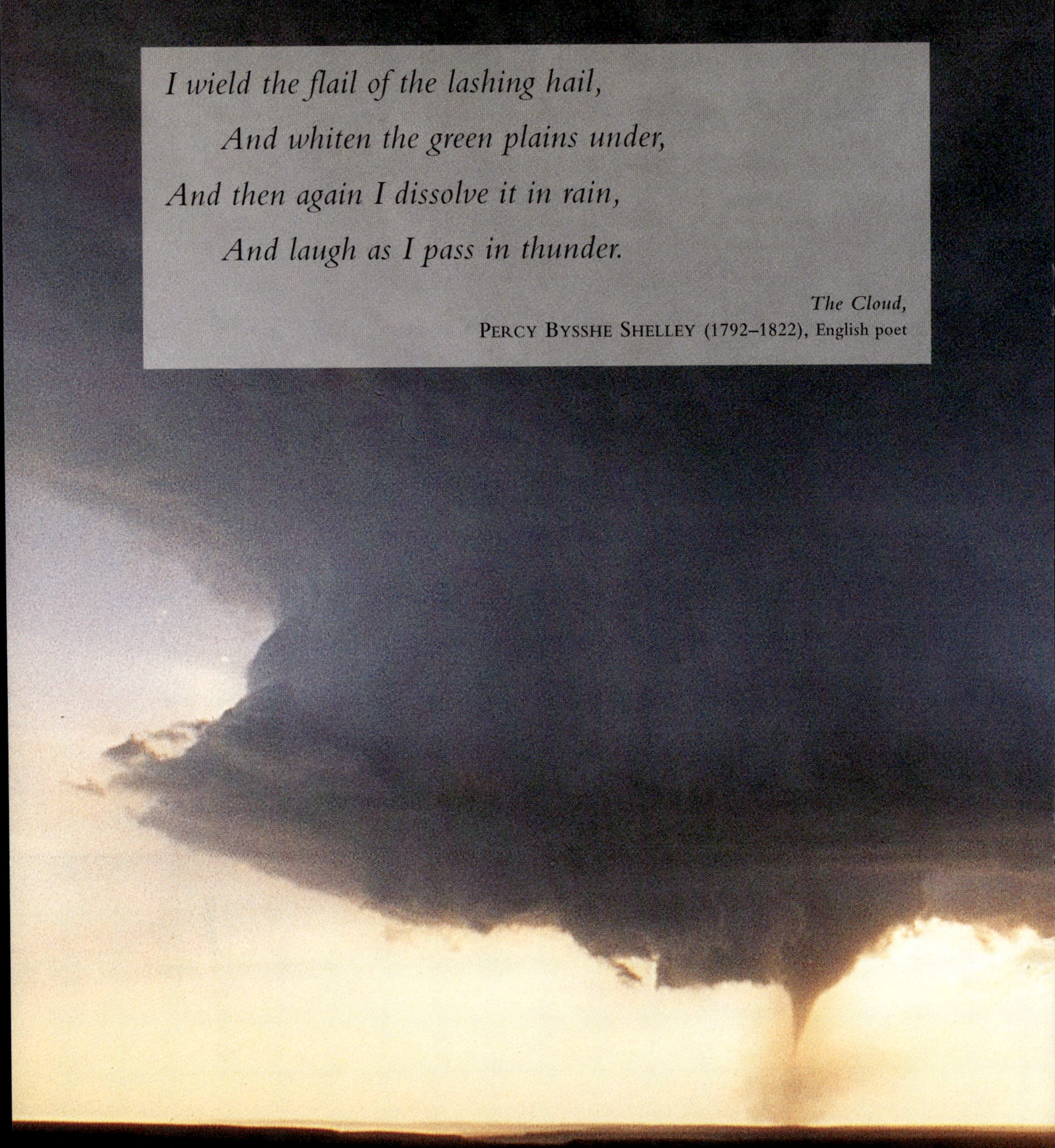

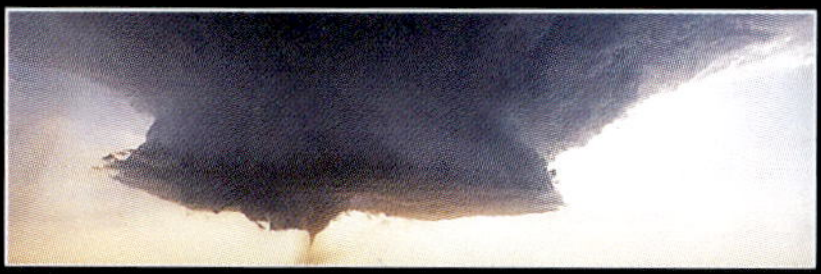

CHAPTER TWO

UNDERSTANDING *the* WEATHER

THE ATMOSPHERE

The Earth is surrounded by a blanket of gaseous chemicals. This blanket, which we call the atmosphere, provides the raw materials for weather and sustains all forms of life on our planet.

Our atmosphere provides oxygen, without which living things would not survive, and contains water vapor, the moisture needed to create weather. The circulations of the atmosphere give rise to winds and storms, and shape our varied climates.

STUDYING THE SKY

The atmosphere is all that protects us from the heat of the Sun and the potentially devastating effects of meteors. Yet, in relation to the diameter of the Earth, the atmosphere is remarkably thin. If the Earth were the size of a party balloon, the atmosphere would be no thicker than the rubber of the balloon.

It was discovered only relatively recently that this fragile shield is made up of several distinct layers. In the nineteenth century, hot-air balloonists found that the temperature of the atmosphere decreased gradually with altitude. For some time, scientists assumed that the temperature continued to fall all the way to outer space. However, studies carried out by French meteorologist Teisserenc de Bort in 1899 (see p. 73) revealed that the temperature actually stopped decreasing at about 6 miles (10 km) above the ground, and a new layer began.

Subsequent studies have found a total of five distinct atmospheric layers from the surface to outer space, although we still do not know how and why the layers formed. The most significant layer from a meteorological point of view is the troposphere—99 percent of our weather occurs here—though all layers have an influence on climatic conditions on Earth.

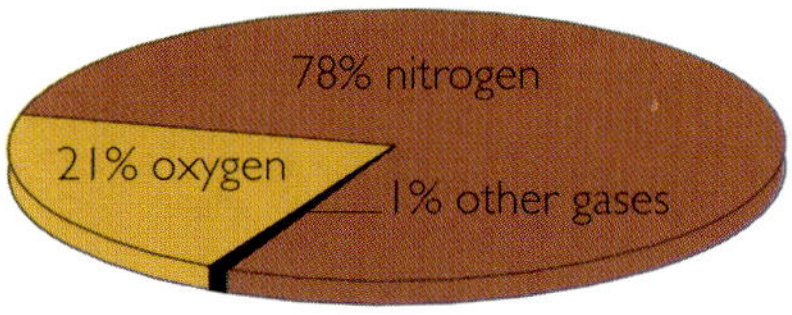

THE AIR WE BREATHE *This pie chart shows the proportions of gases in the troposphere and in the stratosphere. Other gases include argon (0.9 percent), carbon dioxide (0.003 percent), and traces of neon, helium, krypton, hydrogen, and ozone. These proportions differ at higher levels of the atmosphere. In moist air, water vapor comprises 1 to 4 percent, which reduces the percentages of the other components.*

ATMOSPHERIC LAYERS

The troposphere extends from ground level to between 5 and 10 miles (8 to 16 km) above the Earth's surface. The height varies with the amount of solar energy reaching the Earth (see p. 28), being lowest at the poles and highest over the equator. On average, the temperature falls 4° F per 1,000 feet (7° C per km).

The level at which the temperature stops decreasing

THUNDERCLOUDS *seen at sunrise from the orbiting space shuttle* Atlantis. *The flat anvil shapes at the top of the clouds indicate the limit of the troposphere. Much of what we know about our atmosphere was first discovered by intrepid balloonists in the nineteenth century. This print (top left) shows an ascent above Oxford in England.*

with height is called the tropopause, and temperatures here can be as low as -70° F (-58° C). The next layer, the stratosphere, extends to about 30 miles (50 km) above the Earth's surface, and temperatures slowly increase to 40° F (4° C). The ozone layer (see p. 122) is located in the stratosphere, at about 15 miles (24 km) above the Earth. It absorbs most of the Sun's harmful ultraviolet rays.

Above the stratosphere lies the mesosphere. Here temperatures decrease again, falling to as low as -130° F (-90° C). At about 50 miles (80 km) above the Earth, the temperature stops decreasing at the mesopause. In the next layer, the thermosphere (also known as the ionosphere), temperatures rise dramatically, reaching 2,700° F (1,480° C) under certain conditions.

The Sun's rays, including X-rays and ultraviolet radiation, pass into the thermosphere and break the layer's molecules into positive ions and negative electrons. These ions and electrons reflect radio transmissions from Earth back toward the surface, facilitating the reception of long-range radio broadcasts. The thermosphere also protects us from meteors and obsolete satellites, because the high temperatures burn up nearly all the debris coming toward Earth.

In the next layer, the exosphere, there is a variety of gases, including helium, nitrogen, oxygen, and argon. These gases are present only in very small quantities because the lack of gravity at this height allows molecules to escape easily into space. Temperatures here range from about 570° F (300° C) to over 3,000° F (1,650° C).

THE LAYERS OF THE ATMOSPHERE *are defined according to their temperature profile. The colored line on this chart (left) indicates the changes in temperature. In order to show more detail, the proportions of the layers have been distorted. The actual proportions are shown in the blue bands on the right.*

THE SOURCES *of* WEATHER

The interaction of the Sun and the Earth's atmosphere is the main driving force behind weather.

BAROMETERS *are used to measure air pressure. This Italian model (left) dates from the eighteenth century. Low air pressure at high altitudes means that there is less oxygen in the air (above). Differences in pressure give rise to winds (far left).*

The Sun's heat is the origin of all our weather. It causes air masses to form and circulate in our atmosphere. This movement creates differences in air pressure which, in turn, create winds.

AIR PRESSURE

The air is composed of billions of molecules that are constantly moving in all directions, bouncing off whatever they encounter. These collisions constitute what is known as air pressure (or barometric or atmospheric pressure). The more collisions occurring within a certain area, the greater the air pressure.

Although we are unaware of it, air constantly exerts pressure on us—on average, 14¾ pounds per square inch (1 kg per cm^2). Since air molecules are naturally drawn toward the Earth by gravity, the density of air is greater near the surface of the planet. Thus, the air pressure at a certain level, or the number of molecules in a given area, decreases with height. The fact that the air molecules are constantly moving to and fro is what prevents them from settling at ground level.

Air pressure is normally measured in hectopascals (formerly known as millibars) (see p. 96). Typically, the pressure at ground level varies between 980 and 1040 hectopascals, as air rises in some localities and sinks in others.

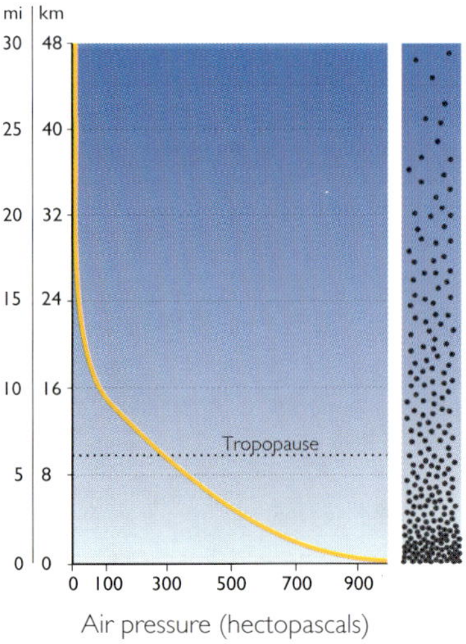

AIR PRESSURE *decreases rapidly with height. This is because gravity pulls air molecules and other atmospheric gases down toward the ground, as shown in the column to the right of the graph.*

CONVECTION

The speed at which air molecules move depends on the air temperature. When an air mass is warmed, the air molecules move faster, causing them to push outward, and the air mass expands. This principle can be demonstrated by cooling or heating a party balloon filled with air and observing how it changes in size—it gets smaller as it is cooled, and expands as it is heated.

The expansion of an air mass causes a drop in density, and the air mass becomes lighter than its surroundings, which causes the air parcel to rise. This process is known as convection. If an air parcel is cooled, the process is reversed and the air tends to sink.

The same process is constantly taking place in our atmosphere, where the Sun

All sorts of things and weather must be taken in together, to make up a year and a sphere.

The Mountain and the Squirrel,
RALPH WALDO EMERSON
(1803–82), American writer

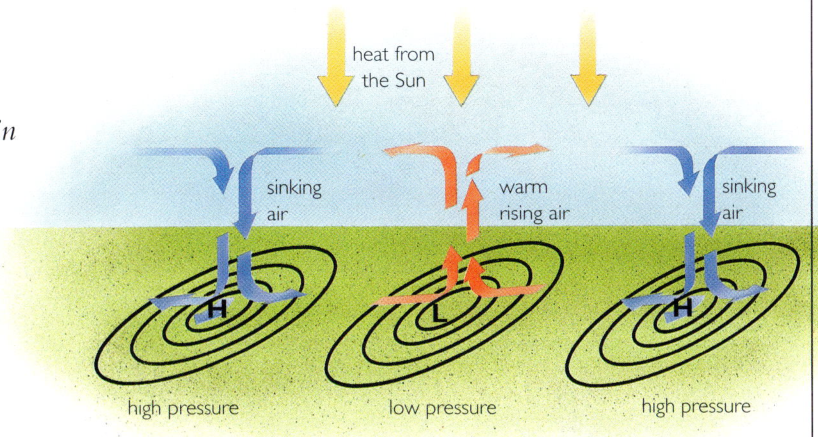

CONVECTION *When warm air rises, it creates an area of low pressure. Eventually, the air cools and sinks toward the ground, forming areas of high pressure.*

provides the heating mechanism. This heating is not uniform and regular, however. Numerous factors affect the amount of solar energy reaching different parts of the globe, including the seasons (see p. 28), latitude, cloud cover, reradiation of heat from the land and the sea, and winds. This means that convection occurs more readily in the warmer parts of the Earth.

CHANGES IN PRESSURE

As a warm air mass rises, it cools and spreads. Once it has cooled, the air starts to sink back to Earth. Where the air is rising, an area of low pressure results; where the air is sinking, high pressure occurs.

Since the atmosphere constantly works to restore equilibrium, air moves into the low-pressure area from surrounding areas of higher pressure. This movement of air from high- to low-pressure areas—the movement is always in this direction—is better known as wind.

The mechanism by which air pressure differences produce motion of air can readily be seen by inflating a party balloon and stopping the air escaping by holding the neck of the balloon. A pocket of "high pressure" is created inside, and when the neck is released, air moves quickly from high to low pressure, equalizing, or balancing, the air-pressure difference.

The difference in air pressure over a horizontal distance is called the pressure gradient force. The greater the difference in pressure between two air masses, the greater the pressure gradient force and the stronger the winds blowing from the high-pressure area toward the low-pressure area.

On a weather map, isobars (see p. 84) are sometimes used to indicate increments in air pressure. The closer the isobars on a map, therefore, the stronger the winds.

AIR PRESSURE AND WEATHER

Often, as air rises and creates an area of low pressure, water vapor in the air will condense and form clouds (see p. 42). Conversely, sinking air generally means that no condensation can take place. Low pressure is therefore generally associated with cloudy skies and wet weather, whereas high pressure is normally associated with clear skies and sunny conditions.

HIGH PRESSURE *usually gives rise to clear skies (right). Rising air creates low pressure and often results in cloud formation (far right).*

Seasons *and* Other Cycles

Our weather is strongly influenced by daily and seasonal variations in the amount of sunlight reaching the Earth.

SUN GODS, *such as this one from fifteenth-century Italy, reflect a universal recognition of our dependence on the Sun.*

The amount of solar energy reaching a certain part of the Earth determines the season there and, ultimately, has a major influence on the climate. Variations in the amount of energy reaching different parts of the Earth are a result of the Earth's tilt, rotation, and orbit around the Sun.

The Earth's Orbit

The Earth is the third planet in our solar system, located about 93 million miles (149 million km) away from the Sun. Earth years are determined by the time it takes for the Earth to orbit the Sun—just over 365 days. The Earth follows an oval-shaped path, called an ellipse, around the Sun. Because of this, the Earth comes closer to the Sun at certain times of the year. This increases the amount of solar heat the Earth receives but is not the cause of the seasons.

HALF LIGHT *The angle at which the Sun strikes the Earth (above) is the key factor in seasonal changes in climate. Certain natural phenomena, such as the fall colors of the leaves of deciduous trees (right), indicate the arrival of a new season in middle latitudes.*

Night and Day

An imaginary line, or axis, runs through the Earth from the North Pole to the South Pole; the Earth spins on this axis, like a spinning top. It takes about 24 hours, or a solar day, for the Earth to spin around once, and this creates day and night. The resulting diurnal changes in the weather, such as high and low temperatures, cause many of our weather patterns.

The Seasons

The Earth is tilted about 23.5 degrees on its axis, and this tilt is the cause of our seasons. As the Earth orbits the Sun, different parts of the planet are tilted toward the Sun, so varying amounts of heat occur around the world at different times of the year.

In the Northern Hemisphere, the North Pole is tilted away from the Sun in December. Less light reaches the hemisphere, resulting in low temperatures and short days—in other words, winter.

As the Earth moves in its orbit, the North Pole begins to tilt toward the Sun. The Sun rises higher in the sky and daylight increases. In March at the vernal, or spring, equinox, day and night are equal.

The amount of sunlight increases until the summer solstice on 21 June, when the Sun reaches its maximum height in the sky. On this day, the Sun is directly overhead at the Tropic of Cancer (23.5 degrees north latitude) and areas north of the Arctic Circle (66.5 degrees north latitude) experience 24 hours of daylight. As the Earth continues its orbit around the

Sun, the North Pole begins to tilt away again. The autumnal equinox occurs in September, and daylight continues to decrease until the winter solstice on 21 December.

In the Southern Hemisphere, the seasons are the opposite. When the summer solstice occurs in the Northern Hemisphere, the winter solstice occurs in the Southern Hemisphere, and areas south of the the Antarctic Circle (66.5 degrees south latitude) have 24 hours of darkness. 21 December, the northern winter solstice, is the southern summer solstice. On this day, the Sun is directly overhead at the Tropic of Capricorn (23.5 degrees south latitude).

THE FOUR SEASONS *are associated with distinct weather patterns. In temperate regions, winter is associated with low temperatures and snow, as depicted in this fifteenth-century illumination, painted for the Duc de Berry.*

In the Southern Hemisphere, the spring equinox occurs in September and the autumnal equinox in March.

In the Tropics

The four seasons occur in the middle and high latitudes of the Earth, where the greatest changes in sunlight and heating occur. In the tropics (that is, between the tropics of Cancer and Capricorn), where the length of day and the amount of sunlight vary much less, only two types of season occur: wet and dry. These too are a result of the Earth's orbit around the Sun.

Where the Sun is directly overhead, convection (see p. 26), storm activity, and precipitation are at a maximum. As this area of intense heat shifts from the Tropic of Cancer (in June) to the Tropic of Capricorn (in December), maximum rainfall moves with it. Tropical areas therefore experience a distinct annual cycle of wet and dry seasons.

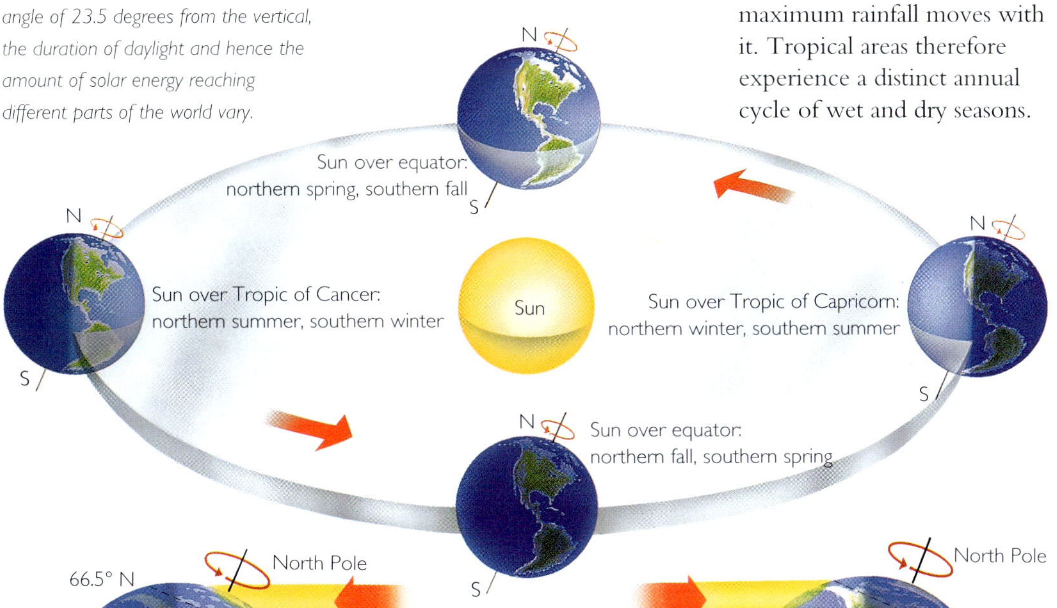

THE EARTH ORBITS THE SUN *once a year. As the Earth is tilted at an angle of 23.5 degrees from the vertical, the duration of daylight and hence the amount of solar energy reaching different parts of the world vary.*

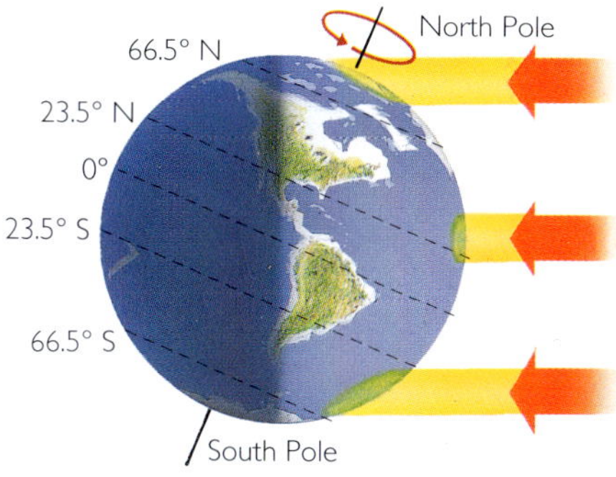

NORTHERN SUMMER *and southern winter occur when the Northern Hemisphere is tilted toward the Sun.*

SOLAR HEATING *is most intense between the tropics, where the Sun is almost directly overhead.*

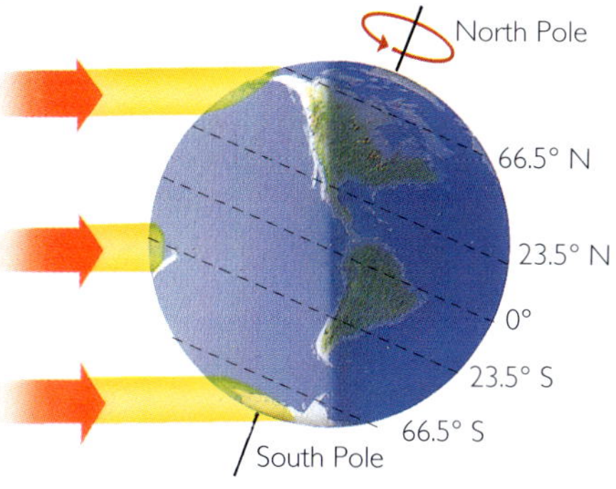

NORTHERN WINTER *and southern summer occur when the Northern Hemisphere is tilted away from the Sun.*

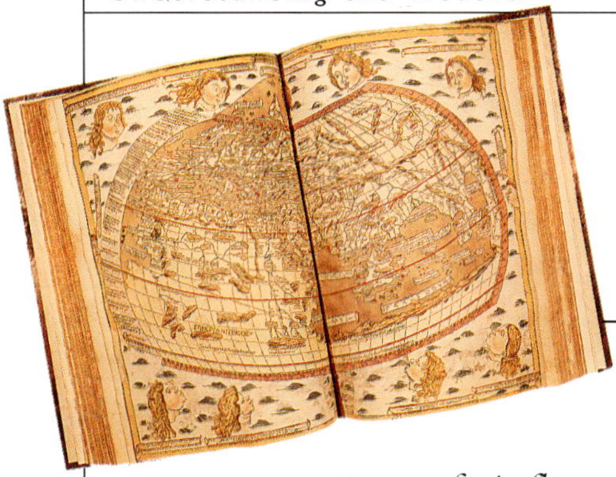

Global Wind Patterns

Uneven solar heating of the Earth creates varying patterns of air flow and, hence, varying weather at different latitudes.

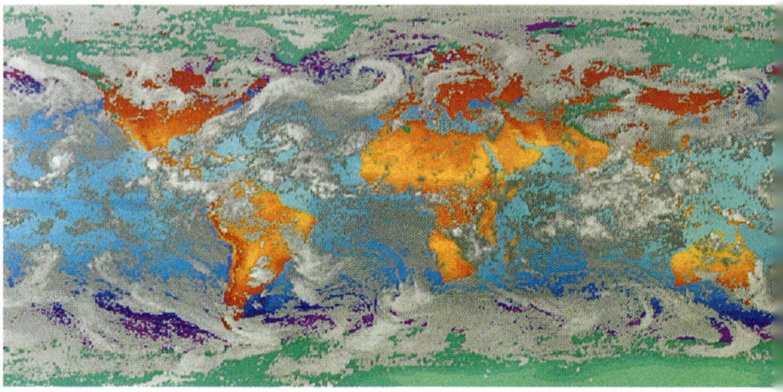

THIS SATELLITE IMAGE *shows how global air flow creates clear skies at around 30 degrees north and south, and unsettled weather at the equator and in higher latitudes.*

The intense heat that reaches the tropics throughout the year produces powerful convection (see p. 26) in these areas. Warm, humid air rises, creating a belt of low pressure, clouds, and rain around the globe at the equator.

The air that rises at the equator eventually meets the tropopause (see p. 24) where it can rise no farther. It then spreads outward toward the poles, gradually cooling and sinking back down to the surface at about 30 degrees north and south latitudes. This sinking air causes an increase in air pressure, bringing generally fair, dry conditions. Most of the world's deserts are located in these high-pressure areas.

Wind Cells

Some of the air from these areas at 30 degrees north and south, forced outward by sinking air, moves back toward the low pressure at the equator. This air flow is known as the trade winds. The area at the equator where these winds die out was named the doldrums (from an old English word meaning dull) by early mariners who feared being stranded there.

WORLD WINDS *Ptolemy's classification of the world's wind regimes is illustrated on the border of this fifteenth-century map (above). The actual pattern (right) is somewhat more complicated.*

Thus, there are circulations of air that rise in the tropics, sink at 30 degrees north and south, and flow back to the equator. These are known as Hadley cells, after George Hadley (1686–1768), the English scientist who first described them in 1753.

While most of the warm air that sinks to the surface at 30 degrees north and south

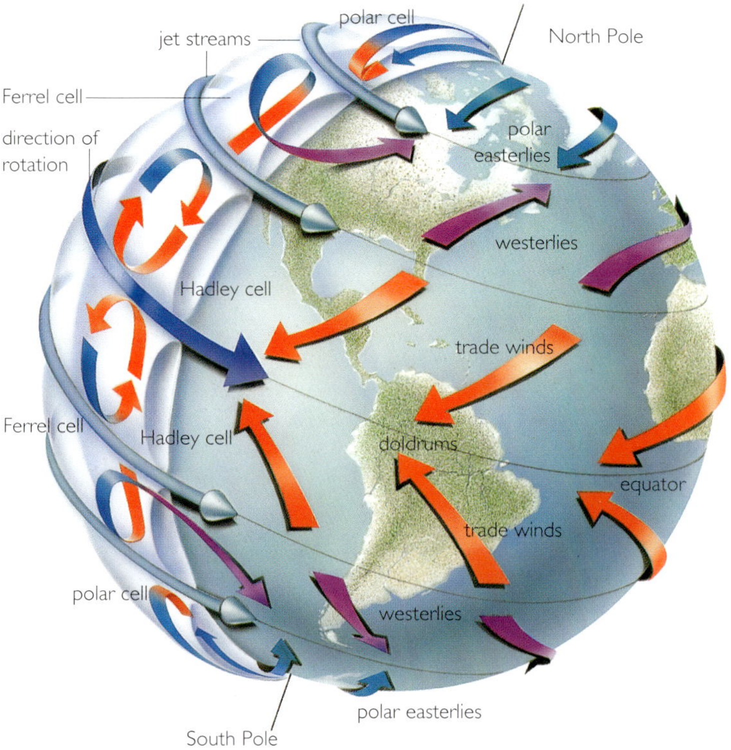

returns toward the equator, some of it continues to move poleward. At approximately 60 degrees north and south, this air meets cold polar air. The areas where these air masses meet are known as the polar fronts. The difference in temperature between the two air masses causes the warmer air to rise. Most of this air moves back toward the equator, sinking to the ground at about 30 degrees north and south and contributing to the high pressure in these regions.

These circulations that occur between 30 and 60 degrees north and south are named Ferrel cells, after William Ferrel, who first identified them in 1856.

The rest of the air that rises at the polar fronts continues to move poleward. As it nears the poles, it cools and sinks, and returns toward 60 degrees north and south. These polar Hadley cells, as they are known, are weaker than the tropical ones because less solar energy reaches polar regions.

The Coriolis Effect

These air flows do not move in a straight north–south path. This is because the Earth's rotation causes any freely moving object or fluid to appear to turn to the right of the direction of motion in the Northern Hemisphere and to the left in the Southern Hemisphere. This effect is known as the Coriolis effect and it was first identified by Gustave-Gaspard de Coriolis (1792–1843) in 1835.

The Coriolis effect explains the flow of weather systems. It causes winds to travel clockwise around high-pressure systems in the Northern Hemisphere, and counterclockwise in the Southern Hemisphere. Low-pressure winds travel in the opposite directions (see p. 34).

JET STREAMS

At high levels of the atmosphere, strong winds develop as a result of significant temperature and pressure differences. These winds, known as jet streams, are usually located at about 30,000 to 35,000 feet (9,000 to 10,500 m), and their speed can be as much as 180 miles per hour (300 kph). Jet streams can strengthen and steer low-pressure systems. During winter, when there are greater temperature contrasts, the jets are more pronounced and shift toward the equator. In summer, when temperatures are more uniform, the jets tend to weaken and shift poleward.

JET-STREAM CLOUDS *(above) seen above Egypt and the Red Sea. The Red Sea is at the top of the picture and the Nile is in the center.*

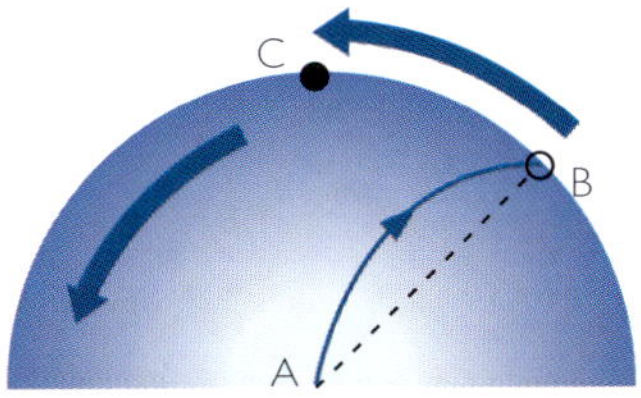

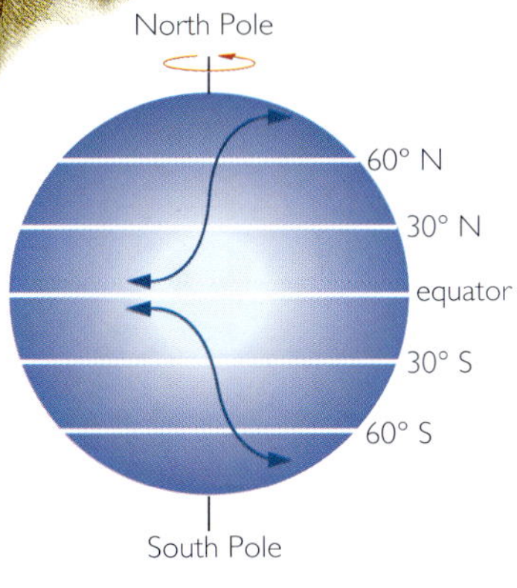

THE CORIOLIS EFFECT *was first explained in 1835 by Gustave-Gaspard de Coriolis (above right). It is best understood by imagining that someone sitting at the center of a moving roundabout (point A, above) throws a ball to someone sitting at a point on the rim (B). By the time the ball reaches B, the person on the rim will have moved to position C. To this person, the ball will appear to have curved away from them. Similarly, to us on our spinning planet, freely moving objects appear to follow a curved path. The result of this (right) is that objects (including weather systems) turn to the right in the Northern Hemisphere and to the left in the Southern Hemisphere. The Coriolis effect does not occur at the equator and is most pronounced at the poles.*

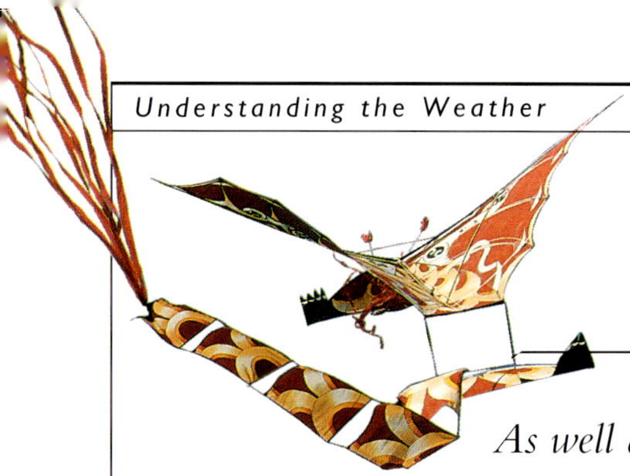

Winds

As well as the major wind systems, there are smaller-scale winds that cause localized weather patterns.

Global air-flow patterns give rise to the principal wind systems, such as the trade winds, the westerlies, and the polar easterlies (see p. 30). They also influence some smaller scale winds such as monsoons.

Monsoon Seasons

Many areas experience monsoons, including the southwestern United States and Chile, but the strongest monsoons occur in southern Asia, northern Australia, and Africa. Monsoons bring copious amounts of rain and can cause massive flooding. Devastating floods have killed thousands of people in Bangladesh, India, and Southeast Asia.

The most dramatic monsoon occurs in India. In winter, when the Sun is relatively low in the sky, the air over Siberia (north of the Tibetan Plateau) cools dramatically, producing strong high pressure. This, in turn, creates winds that blow southeast over India and out to sea, dissipating clouds and rain.

In summer, this high pressure weakens significantly and an area of low pressure develops over northern India. This draws warm, moist air in from the Indian Ocean, which, in turn, produces heavy rain.

THE CHINOOK, *a famous downslope wind that occurs in North America, has created this unusual cloud formation (left). A Chinese dragon kite (above).*

When the moist air reaches the Himalayas, more precipitation occurs as the mountains lift the air. By the time the air reaches the northern side of the range, it has dried out and a rain shadow results (see p. 36). Overall, the monsoon resembles a large-scale version of a sea breeze in summer and a land breeze in winter (see below).

Similar wind shifts, or monsoons, occur in other areas. In the southwestern United States, for example, a monsoon that occurs in late summer and early fall brings moist air in from the Gulf of California. This produces heavy rainfall, which can cause flash flooding in desert areas.

THE SUMMER MONSOON *in India occurs when low pressure over the Tibetan Plateau draws in warm, moist air from the sea.*

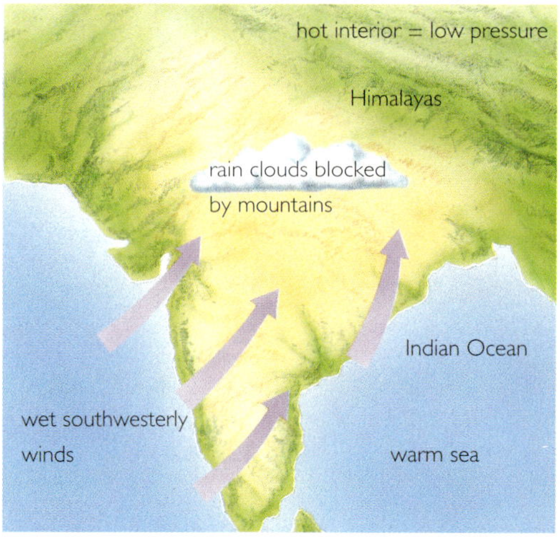

IN WINTER, *intense high pressure in Siberia creates strong northeast winds that keep moist air out over the ocean.*

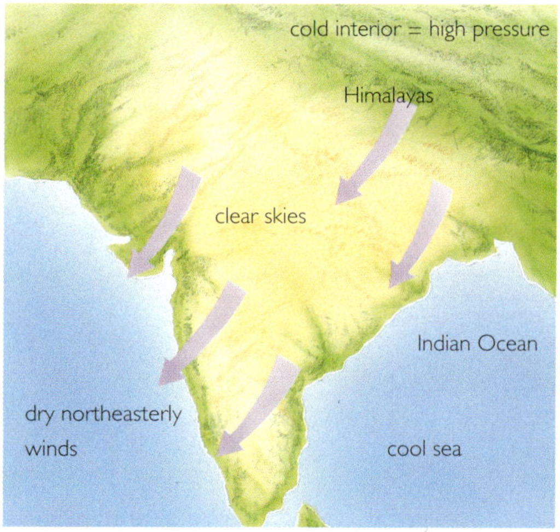

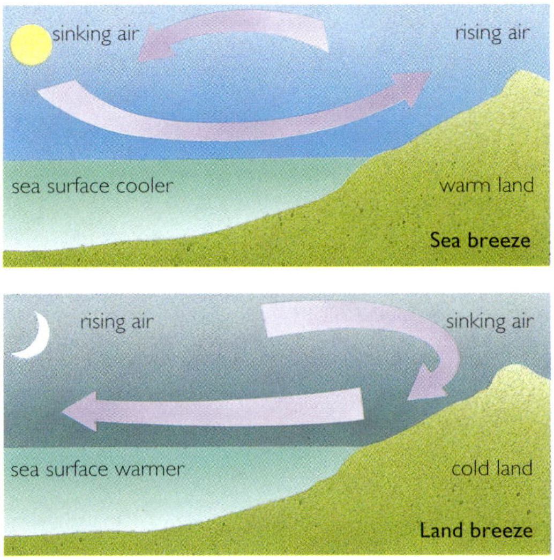

SEA AND LAND BREEZES *occur as a result of temperature differences between coastal waters and the adjacent landmass.*

OVER INDIA *In this satellite image of India, a sea breeze has suppressed cloud development around the coast.*

The oldest voice in the world is the wind.

DONALD CULROSS PEATTIE (1898–1964),
American author and biologist

LOCAL WINDS

Small-scale winds occur as a result of localized differences in pressure or temperature, or the interaction of large-scale winds with local landmasses.

In coastal areas, for example, local winds may develop on clear, sunny days. As the Sun heats the land, the land heats up faster than does the water. The air over the land rises and is replaced by the cooler air from the sea. This circulation is called a sea breeze, and it generally occurs in spring and summer, when differences in temperature between land and sea are most pronounced. The reverse occurs at night: the land cools down quickly, while the air over the sea remains warmer and rises; the air over the land is pushed out to sea, creating a land breeze.

FAMOUS WINDS

As wind blows over mountains and sinks down the other side, it creates high pressure and clear skies. This compression of the air also raises its temperature, resulting in a warm wind. Several winds around the world are examples of warm downslope winds, including the chinook on the east side of the Rocky Mountains and the foehn in Switzerland. These winds can rapidly melt snow and enhance the rain-shadow effect (see p. 36).

Wind forced through valleys will strengthen, just as narrowing the end of a hose will create a more powerful jet of water. In the south of France, a wind formed by the Rhône Valley, known as the mistral, brings cold, dry, and squally conditions from the north.

Other winds result when intense heating of inland areas creates low pressure. A famous example is the sirocco. This brings hot, dry winds to the Mediterranean from the Sahara Desert. These winds pick up moisture from the sea, and, by the time they reach Europe, they are warm and humid. The khamsin, which also originates in the Sahara, brings hot, dry air to southern Egypt, and often devastates crops.

MONSOON FLOODS *occur regularly in northern Indian cities such as Calcutta.*

Frontal Systems

Everyday changes in the weather, from fine, tranquil conditions to very violent storms, are caused by the interaction of different air masses.

THE FIRST SIGN *of a frontal system may be a gradual build-up of cirrus clouds (left). However, cold fronts in particular are often much more sharply defined (above).*

As air masses are moved around the globe by winds, they create weather. When an air mass arrives in a region, it can displace the existing air mass. The boundaries, or leading edges, between different air masses are called fronts: when cold air replaces warm air across a region, a cold front has moved through; a warm front occurs when warm air rides over an existing cold air mass. The interaction of warm and cold air masses may produce low-pressure systems that give rise to unsettled weather, particularly in middle latitudes.

Warm Fronts

When a warm front moves into an area of cold air, it gradually rises over the cold air and cools. Condensation may then follow, resulting in cloud formation. The first clouds to appear ahead of the front are usually cirrus clouds. These are normally followed by a layer of middle-level clouds and then thick stratus clouds. These low-level clouds usually produce widespread precipitation and may be accompanied by fairly strong winds. This situation may last up to a day.

Cold Fronts

Generally associated with low-pressure systems, cold fronts tend to produce more volatile weather than do warm fronts. When a cold front moves into an area of warm air, the warm air, being less dense, is forced sharply upward by the cold air, creating instability and powerful convection (see p. 26). Large cumulus and even cumulonimbus clouds may form, triggering storms along the front. This creates an area of low pressure, which strengthens winds. Rainfall will be heaviest and winds strongest along the front, but showers will also follow as clouds form in its wake.

Cyclogenesis

Low-pressure cells occur when cold and warm air interact to form a rotating weather system—a process sometimes known as cyclogenesis.

When the air masses meet, the warm air rises, creating an area of low pressure where clouds and precipitation develop. The heavier cold air

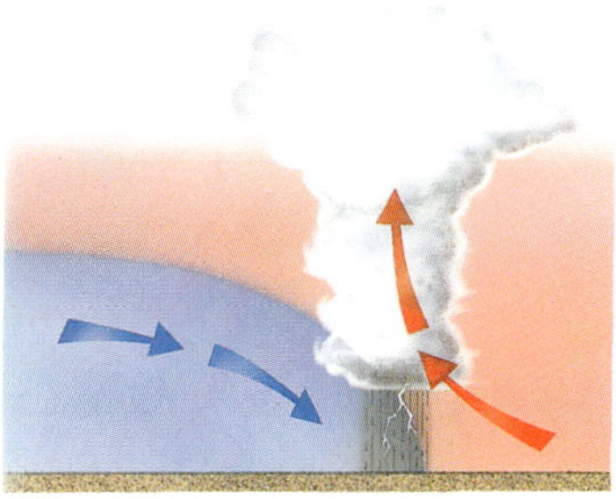

A COLD FRONT *forces warm air to rise rapidly, creating powerful convection that may produce thunderstorms.*

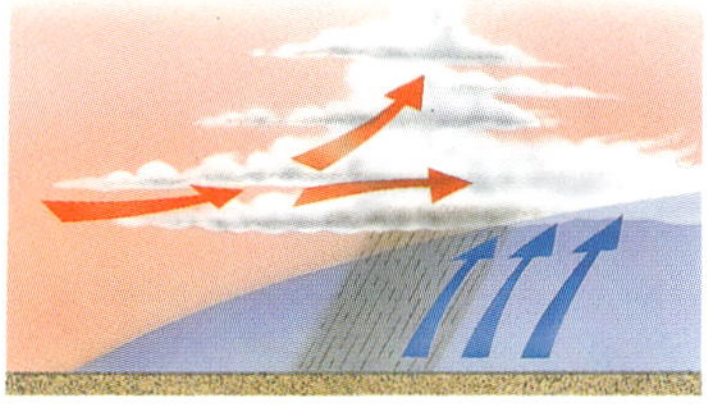

A WARM FRONT *will rise gradually over a layer of cold air, forming cloud that may produce rain over a wide area.*

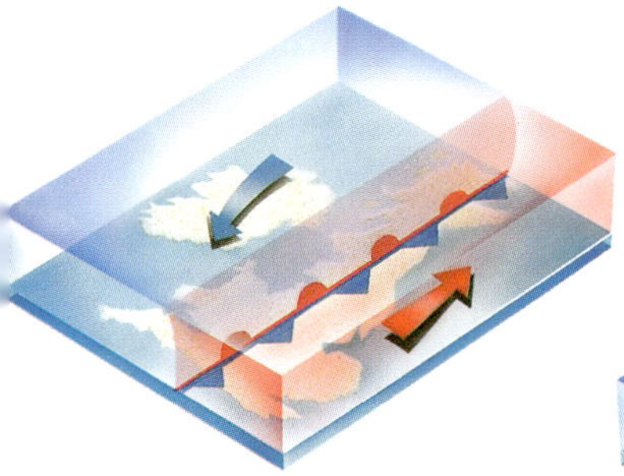

LOW-PRESSURE SYSTEMS *form as a result of the interaction of two air masses of different temperatures.*

◀ ❶ A cold air mass and a warm air mass meet.

OCCLUDED FRONTS *form within low-pressure systems as a result of a cold front overtaking a warm front (see diagram 5, right). This satellite image shows such a front over the British Isles.*

is pulled under it and the faster-moving cold front then begins to catch up with the warm front, forcing more warm air upward.

As air rises and pressure falls, more and more air is pulled into the system and strong winds develop. In the Northern Hemisphere, these winds blow counterclockwise around the low-pressure area; in the Southern Hemisphere, they rotate clockwise.

After about 24 hours, but often more rapidly, the cold front catches up with the warm front. This forms an occluded front, which cuts off the supply of warm air to the system. The air that has risen above the storm gradually cools, the rain stops, the winds drop, and the storm dies out.

ANTICYCLONES

High-pressure systems, or anticyclones, normally result from air sinking and then rotating (clockwise in the Northern Hemisphere and counterclockwise in the Southern Hemisphere). This regularly occurs in warm areas at around 30 degrees north and south latitudes. This type of system is sometimes called a warm high pressure.

High-pressure systems also occur in cold areas, where cold air, being denser than warm air, sinks to the ground, increasing pressure (see p. 26). This often occurs in winter, when less heat reaches the surface of the Earth and dramatic cooling occurs at night, especially if there are no clouds present. This type of high-pressure system is known as a cold high pressure, and is common in inland areas at middle latitudes, such as central Canada and Siberia.

▲ ❷ Gradually, the warm air rises over the cold air. This creates an area of low pressure that the cold front moves into.

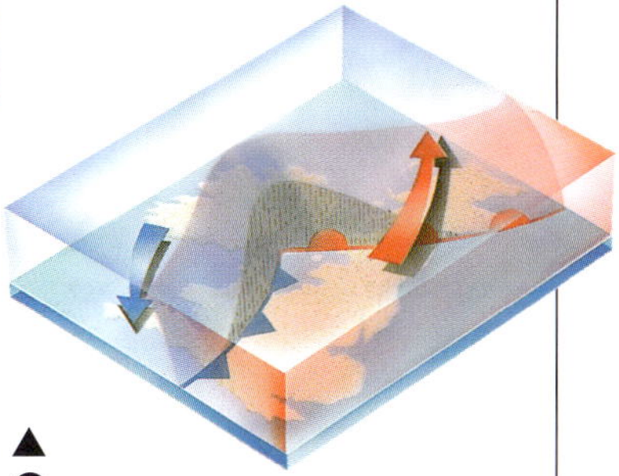

▲

❸ Rising air creates clouds and precipitation, and the fronts begin to rotate.

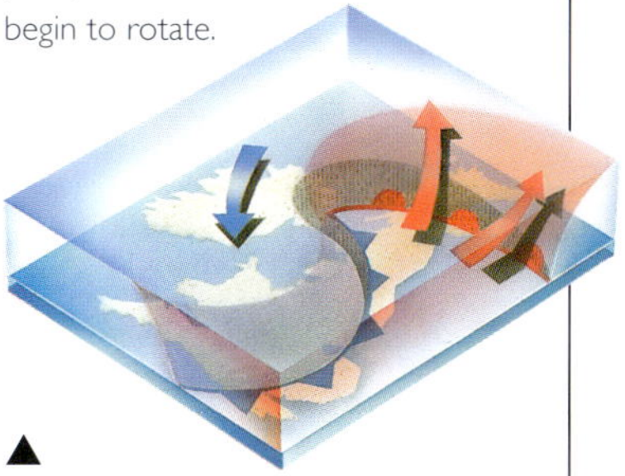

▲

❹ The faster-moving cold front starts to catch up with the warm front. Pressure decreases under the rising air, intensifying precipitation.

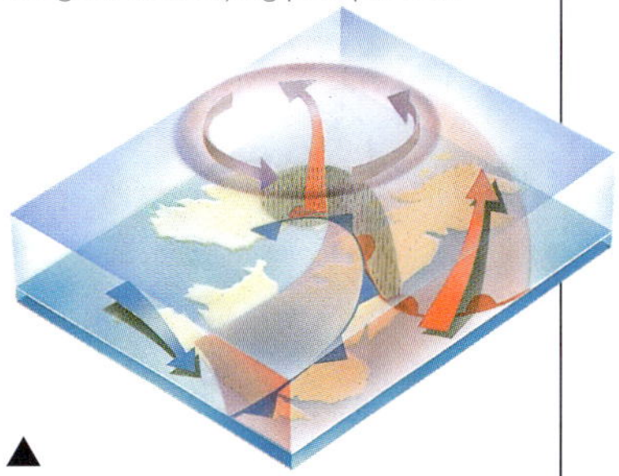

▲

❺ When the cold front catches up with the warm front, an occluded front forms. This creates windy, unsettled weather.

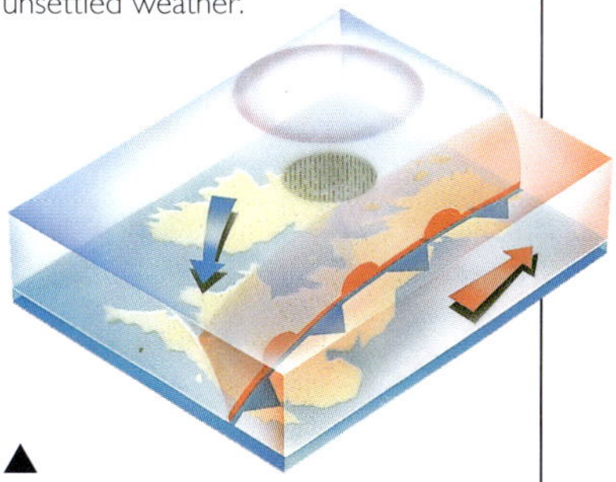

▲

❻ The fully formed occluded front cuts off the supply of warm air, and winds and precipitation subside. If the two air masses subsequently reform, the whole process may begin again.

WEATHER *and the* LAND

Large landmasses can have a dramatic impact on weather patterns around the world.

The differing properties of land and ocean have a highly significant effect on local and large-scale weather systems. The oceans absorb and release heat slowly (see p. 38), whereas landmasses heat up rapidly during the day and cool rapidly at night. There is therefore a greater difference between day and night temperatures in inland areas than in coastal areas. This is why the hottest temperature readings on Earth occur in deserts a great distance from the ocean.

Land also reradiates heat more effectively than sea, so daytime heating leads to widespread convection. Where sufficient moisture is present, cloud will therefore form readily over land.

CONVECTIVE CLOUDS *form more readily over land than over sea (right). A large lenticular cloud over Glacier National Park, Wyoming, USA (below).*

The different heating properties of land and sea play a major role in many wind patterns. Warming of the land north of the Tibetan Plateau is partly responsible for the Indian monsoon, and local temperature differences between land and sea create winds along the coast, known as sea breezes (see p. 32).

LAND AND WEATHER SYSTEMS

If a low-pressure system such as a hurricane (cyclone or typhoon) moves inland from over the ocean, the increased friction caused by rough terrain, mountains, buildings, trees, and so on, will tend to slow the speed of rotation of the weather system, resulting in it losing some of its intensity (see p. 54). However, in some cases, after a hurricane has weakened in this way, it will reintensify as it moves farther inland, forming what is known as an extra-tropical low-pressure system.

MANY ARID AREAS, *such as the Mojave Desert in North America (left), are the result of a rain-shadow effect. The Mojave lies in the rain shadow of western mountain ranges.*

MOUNTAIN WEATHER

Mountain ranges also have a pronounced effect on low-pressure systems. These systems can be thought of as thin, vertical columns of air. The columns contain large amounts of energy, which are manifested in their spin—counterclockwise in the

VALLEY FOG, *as seen in this photograph of the Grand Canyon, Arizona, USA, forms when cold air drains into a valley and cools during the night to its condensation point.*

Northern Hemisphere and clockwise in the Southern Hemisphere. As a low-pressure system approaches the windward side (the side that the wind is coming from) of a mountain chain, it is forced upward and spreads out, which causes the rotation to decrease. When the remains of the low reach the lee side, the column narrows again. The rotation in the column increases and the low re-forms, somewhat weakened.

This weakening and re-formation of low pressure can be compared to the action of a spinning ice skater. With arms extended, the skater spins more slowly, but with the arms either overhead or against the body, the spin increases rapidly.

The Rain-Shadow Effect

Large landmasses have a strong influence on precipitation patterns. In middle latitudes, air generally moves east and so meets the western side of mountain chains. The air is forced upward, causing condensation and hence cloud formation. This process is called orographic lifting and it can produce some unusual cloud types, such as orographic stratus (see p. 192) and lenticular clouds (see p. 210).

Most of the precipitation from the clouds that form over the mountains will fall on the windward side or on the peaks. Depending upon the height of the mountains, great amounts of moisture can be wrung from the air. For example, in an average year in the Sierra Nevada mountain chain of the western United States, more than 100 inches (2,500 mm) of snow and rain fall on the windward side of the range.

On the lee side, as the air sinks back down to lower elevations, a natural warming process takes place, and dramatic changes in temperature and humidity can occur. In some regions, powerful, warm downslope winds are the result. Well-known examples include the chinook and the foehn (see p. 32).

By the time the air sinks to the base of the lee side of the mountains, it is therefore very dry, and an area of extremely low rainfall, known as a rain shadow, is created.

Rain-shadow areas include the high plains of the central United States, east of the Rocky Mountains; and the area east of the Andes in South America. Many of the world's deserts have come about as a result of a prolonged rain-shadow effect.

THE RAIN-SHADOW EFFECT *Mountains cause air to rise, enhancing cloud formation and precipitation. Most of the precipitation will fall on the peaks, and, as the air sinks, the higher pressure will dry it out further, creating arid conditions on the lee side.*

Rising Air

Mountain ranges also tend to trap air. This can lead to the formation of dense fogs. For example, valley fog occurs when cold air drains into a valley and condensation occurs during the night when temperatures fall (see p. 182). Upslope fog forms when warm, moist air rises up a mountain to a level where it cools and condensation occurs (see p. 183).

WEATHER *and the* SEA

Covering more than 70 percent of the Earth's surface, the world's oceans have an enormous effect on climate as they interact with the atmosphere and the land.

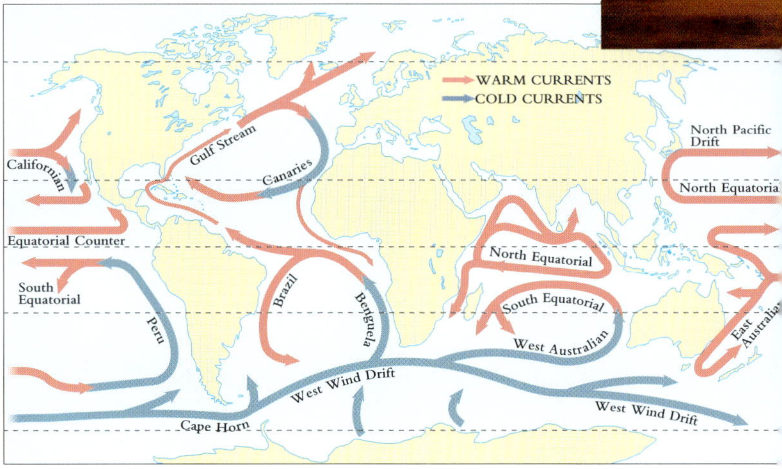

THE PRINCIPAL OCEAN CURRENTS *are influenced by major wind patterns. The currents carry warm and cold water vast distances around the globe.*

The oceans are incredibly efficient storers of heat, and they heat up and cool down much more slowly than do landmasses. This means that ocean currents can carry warm or cold water great distances around the globe. These currents change sea-surface temperatures which, in turn, influence climates, particularly in coastal areas.

On the west coast of the United States, for example, temperatures are moderated by the Californian Current. Especially in summer, onshore winds associated with the current cool the coast, while just a few miles inland, temperatures can rise to more than 100° F (38° C). Similarly, as a result of the warm waters of the Gulf Stream, the winters in Great Britain are very mild compared with those of other countries at the same latitude but outside the current's path.

Ocean currents may also influence rainfall patterns, because warm, moist air associated with warm currents will increase precipitation. For example, variations in the temperatures of ocean currents that originate near Peru may influence rainfall in places as far away as Australia. One manifestation of this process is known as El Niño (see p. 102).

STORMY WEATHER

Ocean currents can play an important part in the development of storm systems. This is particularly the case when cold air from the poles meets warm air associated with warm ocean currents. The interaction of these air masses can produce frontal systems in a process known as cyclogenesis (see p. 34).

Winter storms on the east coast of the United States, for instance, develop and strengthen quickly because of the atmospheric instability arising from the mixture of cold air from the north with the relatively warmer air associated with the Gulf Stream. This process is called explosive cyclogenesis because of the storm's rapid strengthening.

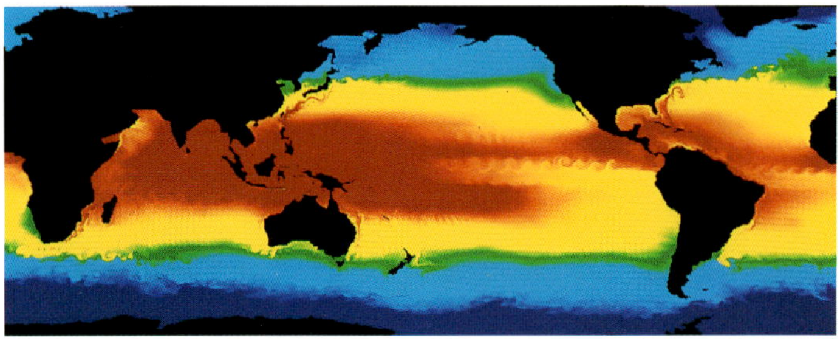

OCEAN TEMPERATURES *In this enhanced satellite image, warm water appears red and cold water blue. The Gulf Stream is evident off the east coast of the United States.*

SEASONAL CHANGES

Because the sea holds heat longer than do continents, it is slower to cool in fall and winter and slower to warm up in spring and summer. There is a lag of several weeks after

COASTAL CLIMATES *are strongly influenced by sea-surface temperatures. The interaction between cold sea air and air warmed by land can create dense sea fogs.*

the winter and summer solstices before the ocean reaches its lowest and highest temperatures of the year.

The resulting differences in temperature between land and sea will have a strong influence on local weather. For example, in spring, coastal areas can be much cooler than inland regions, and sea breezes are common (see p. 32). Similarly, if warm, moist air moves out over colder water, condensation may occur, forming low clouds, fog, or drizzle.

The Water Cycle

A continuous interchange of moisture between the oceans, the land, plants, and clouds fuels much of our weather. This process is known as the hydrologic, or water, cycle.

Studies have revealed that the oceans provide nearly 90 percent of the moisture in our atmosphere. Water leaves the oceans as a result of evaporation (see p. 40). This occurs when the surface of the water is heated: some of the water converts into water vapor and is carried high into the troposphere by rising air.

Moisture also evaporates from rivers, lakes, and other waterways. The rest of the moisture found in the atmosphere is exuded by plants through a form of evaporation known as evapotranspiration.

The water vapor in the air condenses into clouds (see p. 42). These clouds tend to swell and produce precipitation over land, and especially over mountain ranges (see p. 36). Some of the rain and snow that falls is absorbed by plants and soil. The rest runs into rivers or seeps into underground streams. From there, it makes its way back to the sea and the cycle begins again.

Human activities—such as deforestation, the growth of cities, and building dams and reservoirs—affect this cycle by changing precipitation patterns, water storage, and the amount of water that evaporates (see p. 120).

THE WATER CYCLE *is a continuous exchange of moisture between the oceans, the atmosphere, and the land.*

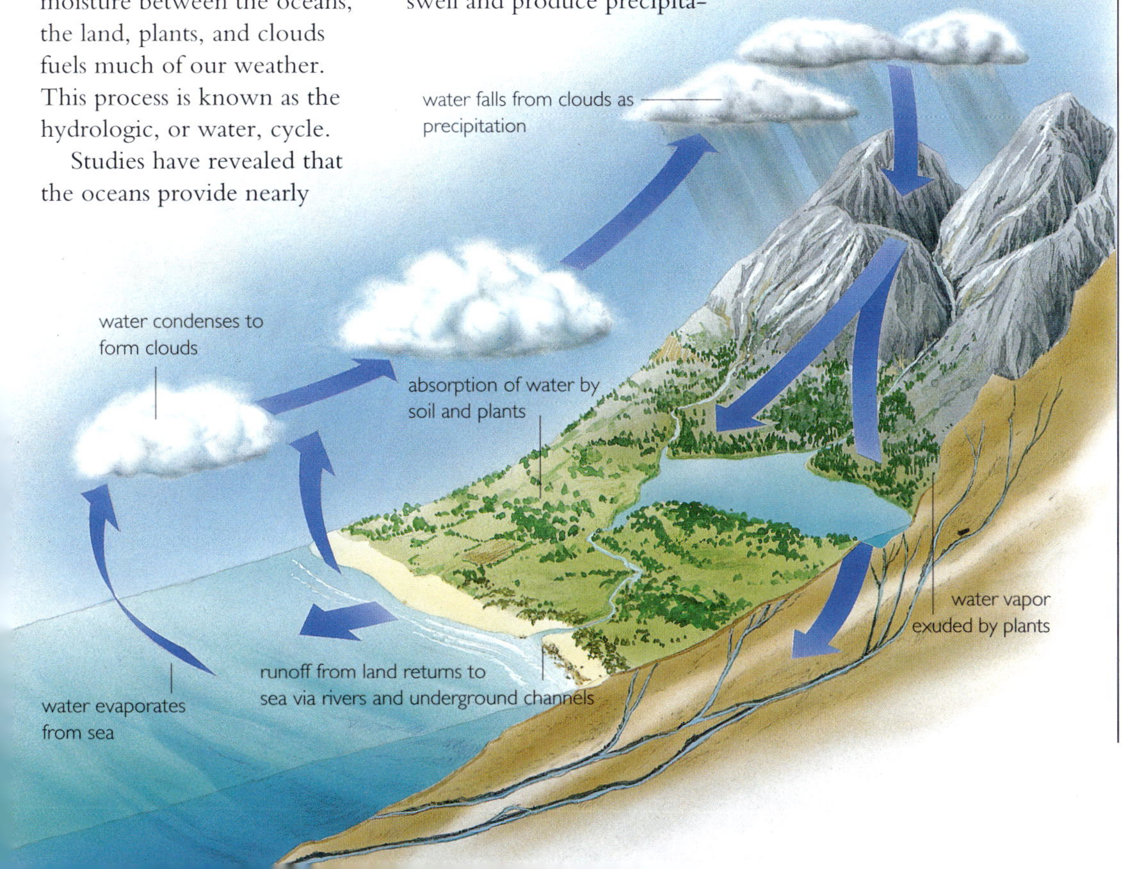

WATER *in the* AIR

Water is an essential element in our weather and exists in many forms, from invisible water vapor to solid lumps of ice.

We are used to seeing water in liquid form or in solid form as ice, but it is also present in the air around us as an invisible gas called water vapor.

EVAPORATION

Around 90 percent of the water vapor in the air comes from the oceans (see p. 39). The water changes from a liquid to a gas (water vapor) through a process called evaporation, which occurs when the Sun heats the water.

Water molecules contain opposing electrical charges, so although the molecules are in motion they remain connected to each other because of the attraction between the charges. Heating causes the molecules in water to move more rapidly. As they speed up, some of the molecules manage to break free and escape into the air above in the form of water vapor. The more water is heated, the greater the amount of water vapor that forms in the air.

WATER *exists in our atmosphere in several forms, including fog—water droplets—(above) and ice (above left).*

CONDENSATION

Air can hold only a certain quantity of water vapor. This amount varies according to the temperature of the air: the warmer the air is, the more water vapor it can hold.

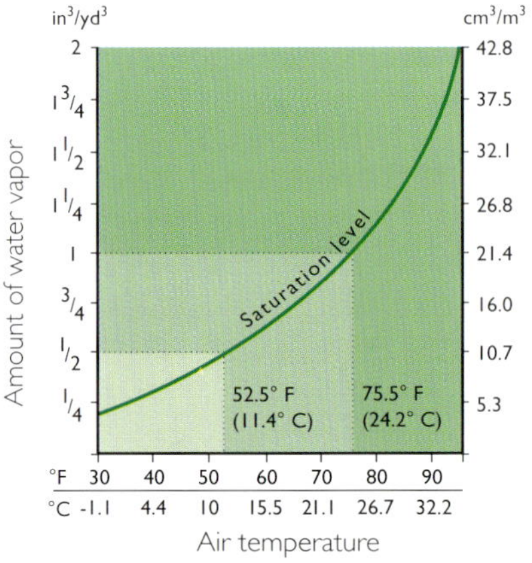

THE AMOUNT OF WATER VAPOR *that an air mass can hold gradually increases as the air temperature rises. When air can hold no more water vapor, it is said to be saturated. The temperature at which saturation occurs is known as the dewpoint. For example, if an air mass holds ½ cubic inch per cubic yard of water vapor (10.7 cm³ per m³), the dewpoint will be 52.5° F (11.4° C).*

When air can hold no more water vapor, it is said to have reached saturation point or be saturated. The water in the air will then begin to condense—that is, form a liquid. The temperature at which water vapor begins to condense is known as the dewpoint.

If condensation occurs at ground level, the water molecules will clump together on surfaces, forming small droplets called dew (see p. 180). When surface temperatures are below freezing—that is, if the dewpoint is below 32° F (0° C)—the water vapor immediately turns, or sublimates, into ice crystals. If dew forms before the temperature falls below freezing, the drops simply freeze. We call both types of ice formation frost, although the first, true frost,

...Under a spring mist

ice and water forgetting

their old differences

TEITOKU MATSUNAGA (1571–1653), Japanese poet

HOAR FROST *on oak leaves and gallnuts (above). Dew covers an argiope spider and its web (below).*

is known more accurately as hoar frost (see p. 186).

Under certain conditions, airborne water vapor droplets may remain in liquid form even though their temperature has dropped below 32° F (0° C). These supercooled droplets, as they are known, will tend to freeze on contact with any surface that has a temperature of below 32° F (0° C). This happens, for example, when supercooled raindrops fall onto frozen surfaces (see p. 222).

MINUTE PARTICLES

If the atmosphere were pristine, condensation could not take place above ground level, because there would be nothing for water vapor to condense onto. However, the air is actually full of microscopic airborne materials, such as sea-salt particles, dust particles, and suspended pollutants. These condensation nuclei, as they are known, allow airborne water vapor to condense.

When condensation takes place just above the ground, fog forms (see pp. 181–5). When it occurs at higher levels, clouds form (see p. 42). The two processes are the same, and fog can be thought of as cloud on the ground.

ABSOLUTE AND RELATIVE HUMIDITY

The amount of water vapor in the air is often expressed as humidity. Absolute humidity is a measure of the volume of water in a certain amount of air at the current temperature. However, since the amount of water vapor that air can hold increases with temperature (see the graph opposite), a more useful measurement is the relative humidity. This is the amount of water vapor in the air expressed as a percentage of the amount required to saturate the air at the current temperature.

Saturated air can be stated as 100 percent relative humidity, while 75 percent relative humidity means the air is holding only three-quarters of its capacity. If the amount of water vapor remains constant, the relative humidity drops as the temperature rises.

Thus, an air mass that has a temperature of 52.5° F (11.4° C) and holds ½ cubic inch of water vapor per cubic yard (10.7 cm^3 per m^3) has a relative humidity of 100 percent. If the temperature rises to 75.5° F (24.2° C), the relative humidity will be 50 percent, which is ½ cubic inch expressed as a percentage of 1 cubic inch per cubic yard (21.4 cm^3 per m^3)—the maximum amount an air mass can hold at that temperature.

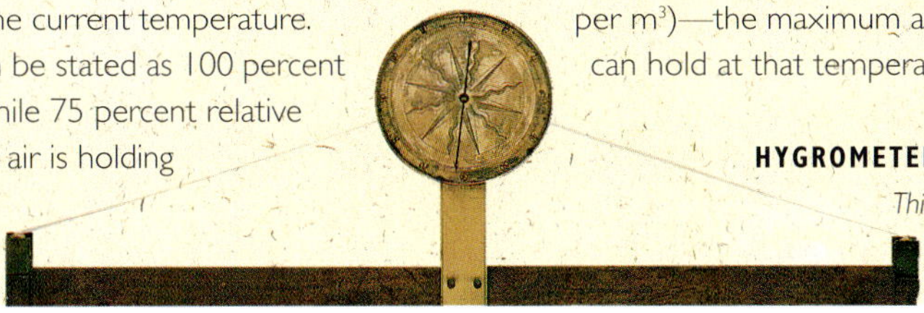

HYGROMETERS *measure humidity. This model dates from the seventeenth century.*

Cloud Formation

A variety of processes can cause air to rise and form clouds of all shapes and sizes.

When condensation (see p. 40) takes place above the surface of the Earth, clouds form. If condensation takes place at a temperature of more than 32° F (0° C), water droplets condense as liquid. If the condensation occurs at a lower temperature, then the water vapor may turn, or sublimate, into ice crystals. Sometimes, water vapor will remain in a liquid, supercooled form at temperatures less than 32° F (0° C) (see p. 40).

Under normal conditions, temperature gradually decreases with height (see p. 24). Clouds forming at high levels of the troposphere therefore tend to be ice-crystal clouds, while those at lower levels normally consist entirely of water droplets. At middle levels, clouds may be made up of a mixture of water droplets, ice crystals, and supercooled droplets.

Hundreds of millions of crystals and droplets together form a cloud. The exact type of cloud depends on factors such as the amount of moisture in the air, the degree of uplift, and the atmospheric stability.

AN INCOMING FRONT *can lift air, triggering condensation. This front over Alaska has formed stratus cloud.*

The Dewpoint

As an air mass rises and cools, it can hold less and less water. Eventually, it reaches its dewpoint (see p. 40), the air mass becomes saturated, and condensation takes place.

The dewpoint and hence the level at which condensation occurs vary according to the amount of moisture in the air. Because air temperature usually decreases with height, the greater the amount of moisture in an air mass, the lower the level at which condensation will occur.

Lifting Mechanisms

There are three common processes that cause air to rise. The first is known as convection (see p. 26). When the ground is heated by the Sun, it reradiates heat into the air above it. Some surfaces, such as paved areas, desert sands, and soil, do this more effectively than others, so parcels, or bubbles, of warm air tend to form over these surfaces. These air parcels rise, and, on reaching their dewpoint, form clouds. The greater the degree of heating, the more powerful the convection.

The second process occurs where fronts form (see p. 34).

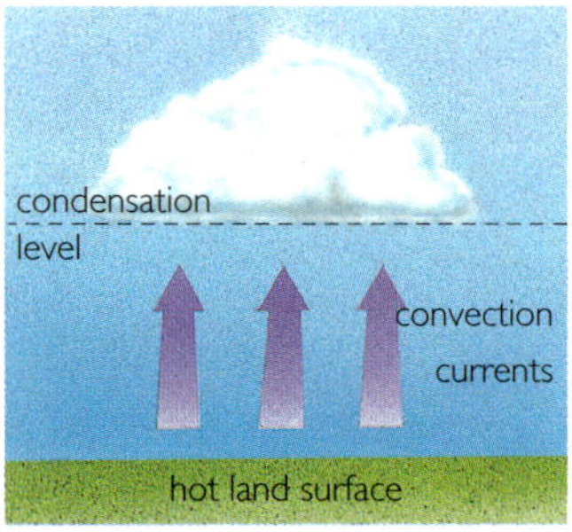

CONVECTION *is the process whereby a hot land surface causes parcels of warm air to rise to condensation level.*

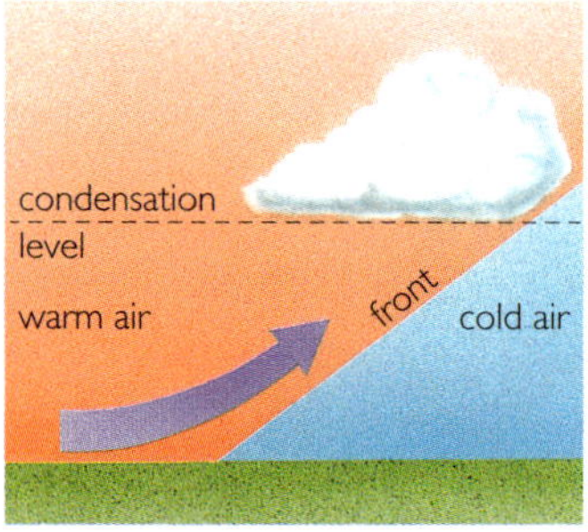

FRONTAL CLOUD *forms when two air masses of different temperatures meet, and the warmer air mass is forced aloft.*

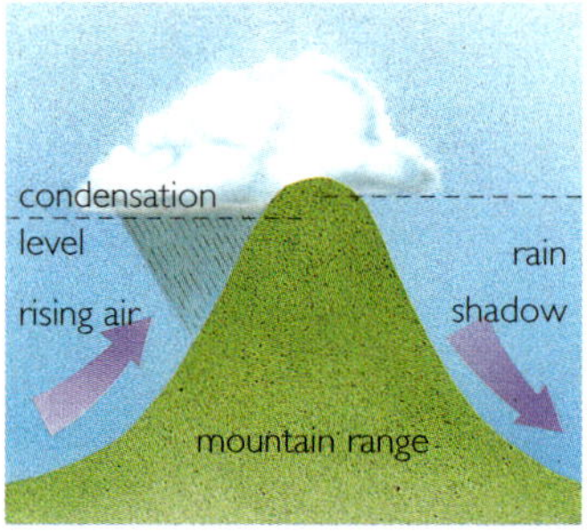

OROGRAPHIC CLOUD *is formed when air is blown up the side of a mountain to its condensation level.*

CUMULUS CLOUDS *(above) result from convection. Lenticular clouds (left) are formed by orographic lifting.*

When two air masses of different temperatures meet, the warmer air will be forced upward by the colder air. If the rising air contains sufficient moisture, cloud will form. This process will produce different types of cloud according to the types of front involved.

The third process, orographic lifting, takes place when an air mass encounters a mountain range. The landmass forces the air upward, often lifting it to condensation level. Orographic lifting can give rise to unusual cloud formations (see pp. 36, 192, 210).

Stability and Latent Heat

An air mass will continue to rise as long as its temperature is higher than that of the surrounding air. If this situation persists as the air mass rises, conditions are said to be unstable. If, however, an air mass quickly reaches the temperature of the surrounding air (and therefore stops rising), conditions are said to be stable.

As long as condensation is not taking place, rising air cools at a rate of 5.4° F for every 1,000 feet (9.8° C per km). Therefore, if we know the temperature of a rising air mass at ground level and the temperature of the air at different levels of the troposphere, we can calculate how far the air will rise (see below).

Generally, it is the temperature of the air at upper levels that determines stability. Cold air above warm air is likely to create instability. Warm air above cold air will generally create stable conditions. This means that it is very important for weather forecasters to know the air temperatures at all levels of the troposphere (see p. 82).

As water vapor in a rising air mass condenses, it releases latent heat. This warms the air mass and increases atmospheric instability, which, in turn, causes the air mass to rise farther. Latent heat is a highly significant factor in the development of thunderstorms (see p. 48).

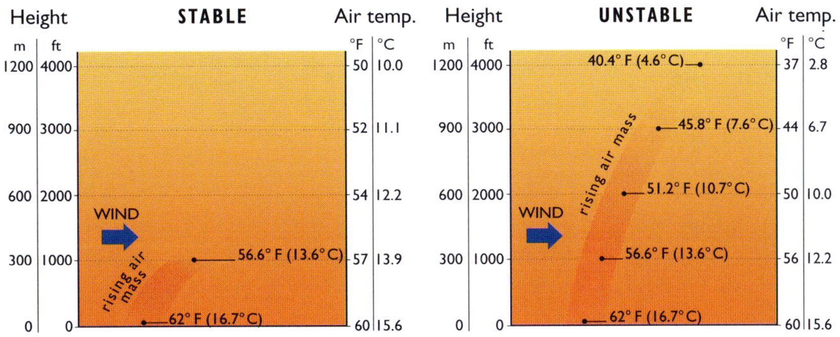

AIR STABILITY *A rising air mass cools at a rate of 5.4° F for every 1,000 feet (9.8° C per km). As long as the air mass remains warmer than the surrounding air, it will continue to rise.*

NAMING *the* CLOUDS

Varying greatly in form and color, clouds are classified according to their shape and their height above the ground.

Fluffy, white shapes like cotton wool drifting in a summer sky, or dark gray towers dwarfing the landscape—clouds in all their forms have delighted observers throughout recorded history. Yet it was not until the early nineteenth century that a system for naming clouds was devised by Luke Howard (see below). Today, all classifications are based on this system.

LOW-LEVEL CLOUDS *include stratocumulus (opposite). A common middle-level formation is altocumulus (above). High-level clouds (top) are known as cirrus.*

GENERAL CATEGORIES

There are two basic types of cloud—cumuliform and stratiform. Cumuliform—from the Latin *cumulus*, meaning heap—are the blossoming, puffy kind. Most often, cumulus clouds are formed by localized convection or orographic lifting (see p. 42). Well-developed cumulus clouds are an indication of unstable conditions.

Stratiform clouds—from the Latin *stratus*, meaning layer—have a flat, layered shape. These result from the uniform lifting of a large, moist air mass, often by a frontal system (see p. 34), and generally indicate a locally stable atmosphere.

CLOUDS AND ALTITUDE

These two types are further categorized according to the height at which they form.

High-level clouds are called cirrus clouds (or have the prefix cirro-) and form above 16,500 feet (5,000 m). Middle-level clouds occur between 6,500 and 16,500 feet (2,000 to 5,000 m) and usually have a name that begins with the prefix alto-. Low-level clouds occur below 6,500 feet (2,000 m). Clouds at low level have no prefix: thus, used on their own, the names stratus and cumulus refer to low-level clouds.

The cumulus and stratus labels are combined with alto- and cirro- to create names for

LUKE HOWARD'S CLOUD CLASSIFICATION

Not until the beginning of the nineteenth century did anyone attempt to apply names to different types of cloud. In 1803, English scientist Luke Howard (1772–1864) presented to his local scientific society a cloud classification based on the most common cloud shapes. To describe these forms, Howard used Latin, the contemporary language of scholarship.

Lumpy, low-level clouds Howard referred to as *cumulus*, meaning heap. More extensive banks of cloud he called *stratus*, meaning layer, while high, wispy clouds he named *cirrus*, meaning lock of hair. The term *nimbus* signified rain-bearing clouds.

The beauty of Howard's classification was that it allowed combinations of these names to be used to identify more specific cloud types. The system won immediate acceptance and admiration. The German poet Johann Wolfgang Goethe (1749–1832) even dedicated four poems to Howard and his system.

Far clouds of feathery gold,

Shaded with deepest purple, gleam

Like islands on a dark blue sea.

Queen Mab,
PERCY BYSSHE SHELLEY (1792–1822),
English poet

middle- and high-level clouds: thus, altostratus means a middle-level layer cloud, and cirrocumulus is a high-level cloud that forms in heaps or lumps.

Cumulonimbus clouds represent a separate category, because they extend from low to high levels. Some classifications incorporate another category, nimbus clouds. However, nimbus simply means rain-bearing—hence, nimbostratus is a rain-bearing, low-level stratiform cloud.

FINE DISTINCTIONS

To further distinguish clouds, additional Latin terms are used to describe other properties. Among the most common are the following:

humilis humble, small
mediocris average, medium-sized
congestus swollen, developing
undulatus undulating, forming in waves
castellanus bearing turrets that resemble battlements
lenticularis lentil-shaped or lens-like
uncinus hooked
fibratus fibrous, forming in strands
nebulosis nebulous, fine

Some of these names are applied to both cumuliform and stratiform clouds; for example, undulatus is used to describe both altocumulus and altostratus clouds that form in parallel bands (see pp. 205, 207). The same term can also be applied to clouds at different levels; thus, undulatus can be used to describe cirrus clouds too, such as cirrostratus undulatus (see p. 215). On the other hand, many of the above names are used for only one type of cloud; for example, humilis always denotes a small, low-level cumulus cloud.

Clearly, this system is not rigid or precise. However, nor are clouds: not only do they form an infinite number of shapes, but they are also constantly changing. To identify clouds, start by assessing the basic type—cumuliform or stratiform—and then try to work out the height. Finer distinctions may be difficult to discern, but you will soon learn to recognize common variations (see pp. 190–217).

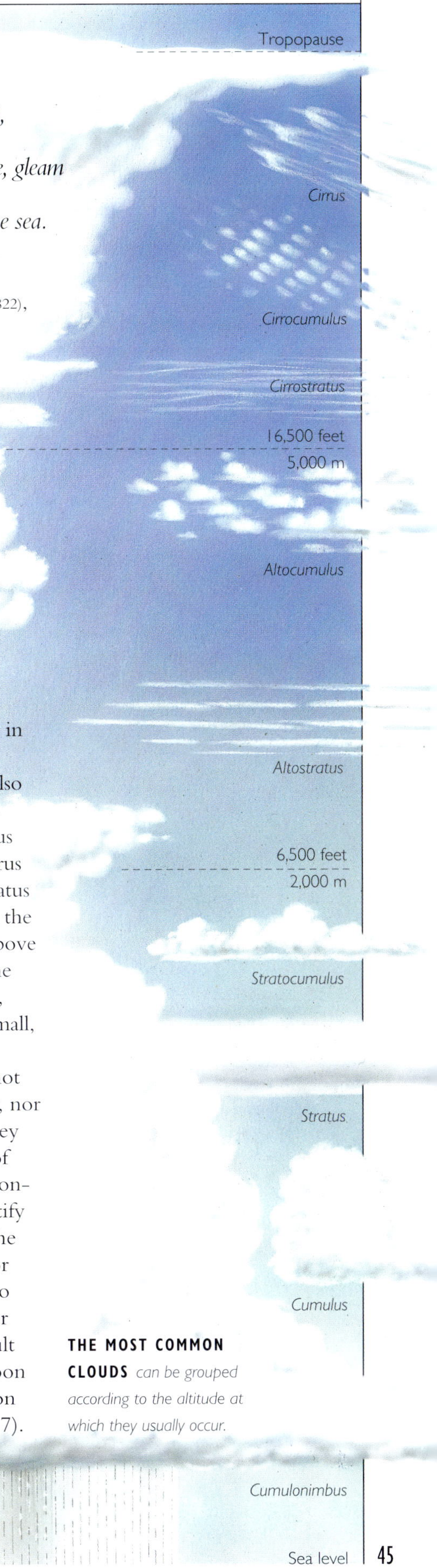

THE MOST COMMON CLOUDS *can be grouped according to the altitude at which they usually occur.*

PRECIPITATION

Rain, snow, and other forms of precipitation result from the build-up of water vapor or ice crystals within a cloud.

Precipitation occurs when some of the millions of tiny water droplets or ice crystals that constitute a cloud grow large enough to fall to Earth under the influence of gravity. Two processes can cause this to happen, and they can occur both individually and in combination.

COALESCENCE AND ICE CRYSTALS

The first process, called coalescence, is most likely to occur in very moist cumuliform clouds where air temperatures are above freezing. The water vapor droplets in clouds are generally so small that they are kept aloft by air resistance and rising air currents, despite the effects of gravity. However, if turbulence within the cloud causes these droplets to collide, they may merge with one another to form larger droplets, and eventually these droplets will be heavy enough to fall from the cloud. As they fall, they collide with more droplets, continuing to grow in size until they reach the ground as rain.

The second process requires ice crystals to be present in the cloud, and is known as the Bergeron–Findeisen process after the Swedish and German scientists who studied it in the 1930s. It is most likely to occur in thick clouds at middle or high latitudes, where supercooled water droplets (water cooled below freezing but not frozen) and ice crystals coexist (see p. 40).

Ice crystals and supercooled water droplets have different saturation points, so when they coexist in a cloud, water molecules will move from the water droplets to the ice crystals. Under these conditions, the ice crystals will grow quickly at the expense of the water droplets until they are large enough to fall. When falling, they will tend to grow further through the process of coalescence, and will melt or remain frozen depending on the temperature of the air below the cloud.

FROZEN PRECIPITATION *can take many different forms, including hail (far left) and snow (left).*

HOW BIG IS A RAINDROP?

The average diameter of a cloud droplet is around 20 microns (1/1200 inch or 0.02 mm). The average diameter of a raindrop is about 2,000 microns (1/12 inch or 2 mm)—100 times larger, as shown in the comparison on the left. This cartoon from 1835 (right) illustrates an early variation on the well-known saying used to refer to very heavy rain, "It's raining cats and dogs."

A SHOWER *falls over the Serengeti in Tanzania, East Africa. Showers from well-developed cumulus clouds may be very heavy.*

Classifying Precipitation

Precipitation is classified according to its form when it reaches the ground. This depends on the formation processes and the air temperatures within and below the cloud. For example, very cold air within a cloud may produce rain, freezing rain, or snow, depending on the temperatures of the layers of air beneath the cloud. Rain may become supercooled and form freezing rain by passing through a layer of freezing air (see p. 222). Ice crystals may melt to become rain (see p. 220), or remain frozen, grow by accretion, and reach the ground as snow (see p. 224). Rain or ice crystals may pass through a warm, dry layer of air and evaporate completely, forming virga (see p. 220).

Rain and Showers

Precipitation that reaches the ground in liquid form, often referred to simply as rain, is defined in a number of ways depending on the size of the water droplets, associated visibility, and the type of cloud from which it fell.

The lightest form is drizzle, which occurs as fine drops falling close together. Mist, being slightly finer than this, does not fall and is therefore considered a form of light fog. Raindrops are larger—from $1/50$ inch (0.5 mm) to $1/4$ inch (6.35 mm) in diameter. Any drops larger than this will normally be broken up by air resistance.

Precipitation may also be classified as steady or intermittent. This depends largely on the type of cloud from which it fell. Generally, steady rain or snow tends to fall from widespread stratiform clouds, such as stratus or altostratus. These usually occur as a result of frontal activity (see p. 34), common in middle latitudes.

Intermittent precipitation, or showers, normally falls from cumuliform clouds, which result from a combination of convection and atmospheric instability (see p. 42).

THE TYPE OF PRECIPITATION *that reaches the ground depends on the process that occurs within the cloud and the temperature of the air between the cloud and the ground. In this diagram, the blue panels under the clouds represent subfreezing air and the red panels represent warmer air.*

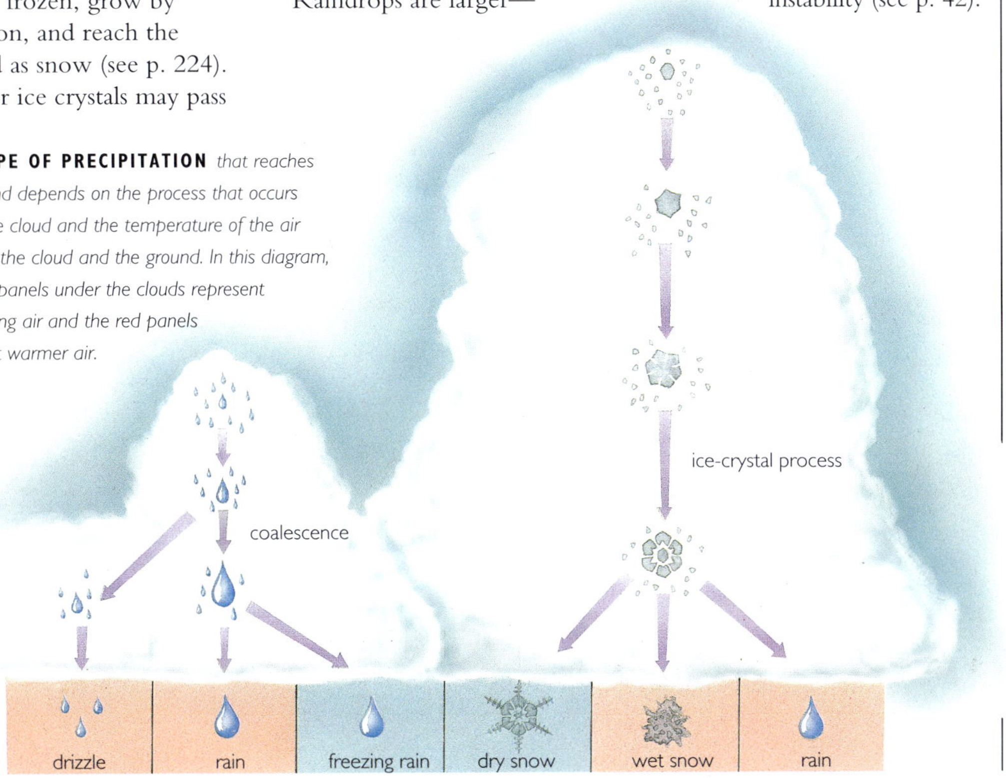

The Life Cycle of a Thunderstorm

White lightning crackling in a purple-black sky, the deafening crash of thunder, and rain or hail lashing down—thunderstorms can be spectacular events.

Every day, about 40,000 thunderstorms occur throughout the world, with the most frequent storms brewing in equatorial regions. The power of a storm can be awesome, as rain, hail, strong winds, or tornadoes are unleashed, accompanied by bright flashes of lightning and the roar of thunder.

Three main ingredients are needed for thunderstorms to form—moisture, instability, and lift—and there are three stages in the formation of a storm: the developing, mature, and dissipating stages.

Gathering Clouds

The developing stage occurs when warm, moist air begins to rise into the normally cooler air above. As the air cools, condensation takes place and clouds form. If convection is strong enough, the clouds continue to develop to the congestus stage (see p. 196).

For the cloud to develop farther, atmospheric instability (see p. 42) at middle and upper levels of the troposphere is required. Latent heat created by the condensation process will enhance instability by warming the rising air.

Once the cloud reaches the cumulonimbus stage (see p. 200), it stops rising only when it meets the tropopause (see p. 24). There, the cloud spreads out in a flat-topped formation that resembles a blacksmith's anvil. At this height, temperatures are well below 32° F (0° C) and the top of the cloud consists of ice crystals.

Sometimes, the updrafts punch through the tropopause into the stratosphere. This is called an overshooting top and may be a sign of a severe or tornadic thunderstorm.

LIFE CYCLE OF A STORM *When warm air rises, condensation occurs and clouds form. If convection is powerful, the cloud may develop to the congestus stage (above). In the mature stage, the top of the cloud spreads out, and cool air descends. When the downdrafts cut off the supply of warm air, the storm dissipates, leaving wisps of cirrus, and small altocumulus clouds.*

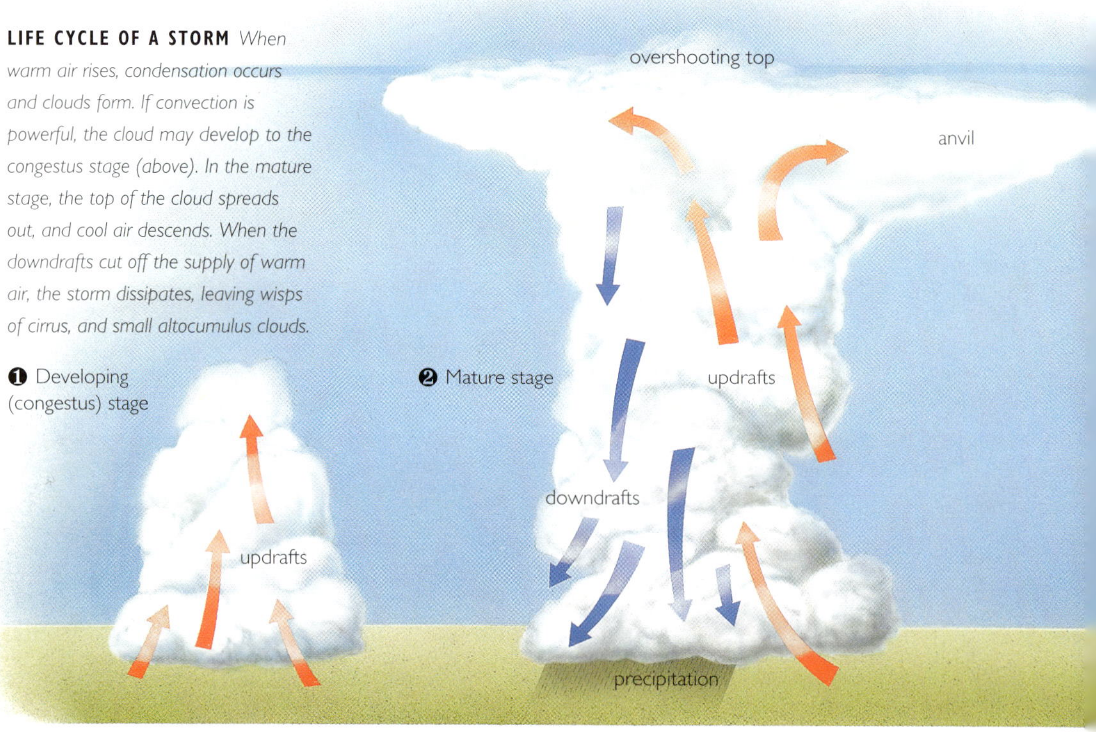

THE ANVIL *of a mature thundercloud (left). As the storm dissipates the anvil may be carried away by upper-level winds (above).*

Stormy Skies

As air cools at the top of the cloud, it starts to descend, aided by gravity and precipitation, causing downdrafts. The cloud has now entered the mature stage, the most destructive phase of a storm.

The upward and downward motions of the air currents enhance the creation of opposite electrical charges that produce lightning (see p. 50). As the lightning's heat passes through the air, the air expands rapidly and creates a sound wave, which we hear as thunder. The electrical charges also enhance coalescence within the cloud (see p. 42), which increases rainfall. Mature storms may produce up to 4 inches (100 mm) of rain per hour and cause floods.

Other by-products of severe thunderstorms include hail (see p. 226), tornadoes (see p. 52), and damaging winds caused by violent downdrafts, known as microbursts (see p. 246).

As the cold downdrafts increase in number and strength, the storm enters its dissipating stage. The downdrafts spread cold air along the ground, and these gust fronts, as they are called, cut off the supply of warm, moist air to the thunderstorm, causing it to weaken. Depending on the type of thunderstorm, its life cycle can last from 15 minutes to several hours.

SQUALL-LINE *storms form along, and are fueled by, a cold front. In this satellite image of the Ivory Coast, a squall line is clearly visible (below).*

Types of Storm

There are different types of storm. Air-mass thunderstorms form as a result of convection alone and are not associated with a frontal system. They can be single- or multi-cell. A single-cell thunderstorm develops and weakens rapidly. Multi-cells are clusters of storms that feed off each other. The downdrafts from each individual thunderstorm spread out on impact with the ground, forcing warm surface air to rise. If other storms are nearby, this rising air fuels their development.

Frontal storms form in a line along a cold front, known as a squall line. These storms are fueled by the front and have an abundance of moisture, lift, and instability. Sometimes, at the end of a squall line, exceptionally strong, self-sustaining thunderstorms can form. Known as supercells, these storms can last for several hours, because the cold front supplies a continuous flow of cooler air at middle levels, which enhances instability. Supercell storms cause the most destructive winds, hail, and tornadoes.

Electrical Phenomena

The dazzling visual display and sound of a thunderstorm are both produced by lightning, an electrical discharge that causes the air to expand.

Lightning is the result of a build-up of opposite electrical charges within a cumulonimbus cloud. Exactly how this happens is not yet clear, but it seems that ice crystals, which form in the upper part of the cloud (see p. 48), are generally positively charged while water droplets, which tend to sink to the bottom of the cloud, are normally negatively charged. It may be that updrafts carry the positive charges up and downdrafts drag the negative charges down.

As this build-up occurs, a positive charge also forms near the ground under the cloud and moves with the cloud.

TYPES OF LIGHTNING *depend on the locations of the opposite electrical charges in and around the cloud.*

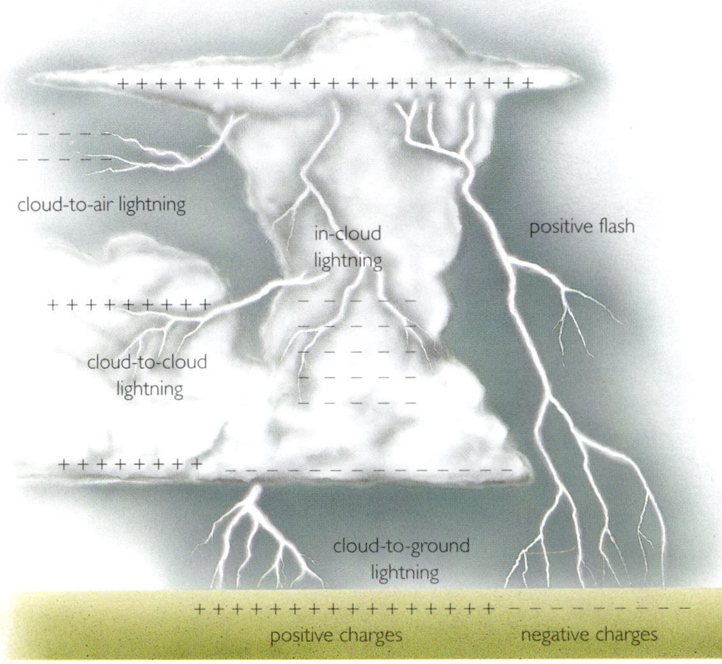

VOLCANIC ERUPTIONS, *such as this one at Mount Kilauea in Hawaii, are often accompanied by lightning.*

Opposites Attract

The opposing electrical charges are strongly attracted to each other. Eventually, the insulating layer of air between the charges cannot keep them apart any longer and a discharge takes place. Negative charges move toward positive charges in an invisible, jagged, zigzag pattern, called a stepped leader. When the negative charge meets the positive charge, a massive electrical current—the lightning bolt—is created and then sustained by a return positive charge back to the cloud. This positive charge travels extremely quickly—about 60,000 miles per second (96,000 kps).

All of this can be repeated rapidly in the same lightning bolt, which gives the lightning its flickering appearance. The process continues until all the charges in the cloud have dissipated.

Most discharges take place within a cloud, between clouds, or between a cloud and the air if there is sufficient charge in the air. Only about one in four lightning bolts strikes the ground. When this happens, the descending leader attracts ground-level positive charges upward, usually from an elevated object such as a tree or building. A lightning bolt that travels all the way from the top of a cloud to negatively charged ground outside the area beneath the cloud is known as a positive flash.

CLOUD-TO-GROUND LIGHTNING *strikes the rim of the Grand Canyon in Arizona, USA (above). This photograph (right), taken by astronauts on the space shuttle* Discovery, *shows a cumulonimbus lit up by lightning within the cloud.*

All forms of lightning may appear as forked, sheet, or streaked lightning, depending on how far the observer is from the charge (see pp. 240–3).

The Sound and the Fury

The temperature of a lightning bolt exceeds 40,000° F (22,000° C). When a bolt forms, the air around it is superheated, which causes the air to expand then contract rapidly. This creates sound waves that we hear as thunder. Because light waves travel much faster than sound waves, we see the lightning before hearing the thunder. It takes about 5 seconds for thunder to travel 1 mile (3 seconds/km), so it's possible to calculate how far away a storm is by counting the seconds between seeing the lightning flash and hearing the thunder, and dividing by five, for miles (or three, for km). Generally, thunder is inaudible farther than 20 miles (32 km) away.

September lightning ...

White calligraphy

On high

Silhouettes the hill

Joso (1662–1704), Japanese poet

Sparks May Fly

There are two other types of atmospheric electricity. A rare form, known as ball lightning, occurs when some of the charge from a cloud-to-ground strike forms a small, round ball. This ball of light may roll along the ground or climb objects until it either explodes or dissipates.

Sometimes, when the build-up of opposite charges is insufficient for a lightning bolt to form, a mass of sparks appears high above the ground in the vicinity of the thunderstorm. This phenomenon was first noted at the top of ships' masts and was subsequently named St Elmo's fire, after the patron saint of sailors.

FRANKLIN'S EXPERIMENTS WITH LIGHTNING

Benjamin Franklin (1706–90), an American author, inventor, scientist, and diplomat, proposed a series of experiments, conducted in France in 1752, that proved that lightning was an electrical discharge. Later that year in Philadelphia, Franklin carried out an experiment during a thunderstorm with a kite attached to a wire with a key at the end of it. When he launched the kite, lightning hit it, traveled down the wire, and electrified the key. Franklin was lucky not to be killed.

Several years earlier, Franklin had theorized that a tall, thin, metal rod attached to the top of a roof and connected to a wire that led to the ground outside the building would conduct electricity from lightning bolts and allow it to pass harmlessly to the ground. This invention, presented to the public in 1753 and called the lightning rod, or conductor, subsequently became a standard attachment on buildings in America and Europe.

WHIRLING WINDS

Whirlwinds range from small dust devils to the strongest winds of all—tornadoes.

Tornadic, or whirling, winds include waterspouts, dust devils, and, of course, tornadoes (see pp. 244–8). They are all characterized by a rotating column of air, but are formed in different ways. Dust devils and small whirlwinds are created by intense local heating of the ground causing air to rise, while tornadoes result from the interaction of warm and cold air currents aloft and are always associated with severe thunderstorms.

TORNADO DAMAGE *(right) occurs frequently in North America, especially during the periods indicated below.*

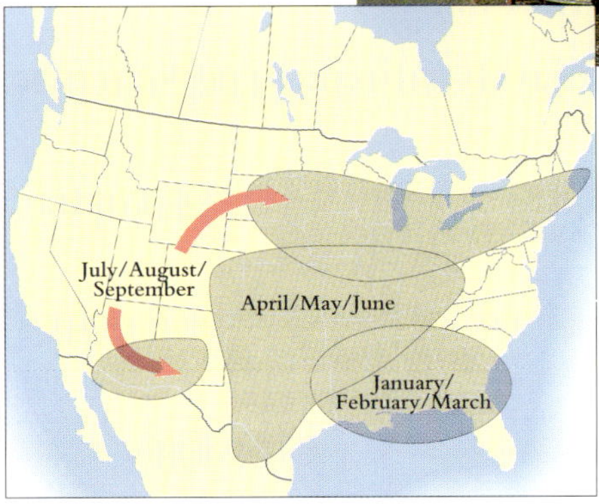

TERRIBLE TORNADOES

Tornadoes are by far the most significant and destructive whirlwinds. They can exceed speeds of 300 miles per hour (483 kph) and wreak havoc on the ground, picking up and hurling large objects such as buildings and cars, and posing a very real threat to life. The deadliest tornado was the Tri-State Tornado, which passed through Missouri, Illinois, and Indiana in the United States on 18 March 1925. In 3½ hours, it killed 695 people along its 219 mile (353 km) path.

Tornadoes are most common in the United States and Australia, but sometimes occur in other areas, including Canada, New Zealand, and the United Kingdom. In the United States, most tornadoes develop in late winter and spring, when cold air moving south from the North Pole and moist, warm air moving north from the tropics meet and fuel atmospheric instability. In Australia, the tornado season lasts from November to May.

WATERSPOUTS *often emerge from congestus clouds. However, they are sometimes associated with severe thunderstorms, in which case they form in the same manner as tornadoes.*

THE FUJITA SCALE

The Fujita scale provides a measure of the strength of a tornado. It was developed by Dr T. Theodore Fujita (b. 1921), of the University of Chicago, who has been studying tornadoes for decades.

Scale Number	*Wind Speed (mph [kph])*	*Damage Type*
F0	40–73 (64–117)	Light
F1	74–112 (118–180)	Moderate
F2	113–157 (181–251)	Considerable
F3	158–206 (252–330)	Severe
F4	207–260 (331–417)	Devastating
F5	more than 261 (418)	Incredible

HOW A TORNADO FORMS *High-speed winds at upper levels of a storm can cause it to rotate (below). The rate of rotation will be much higher at the center of the storm, near the main inflow of warm air. A spinning column of air descends through the updraft area and emerges below the cloud base (see detail, right). As it reaches the ground, a tornado forms. Its fierce, spiraling winds and intense updrafts will destroy everything in its path (bottom).*

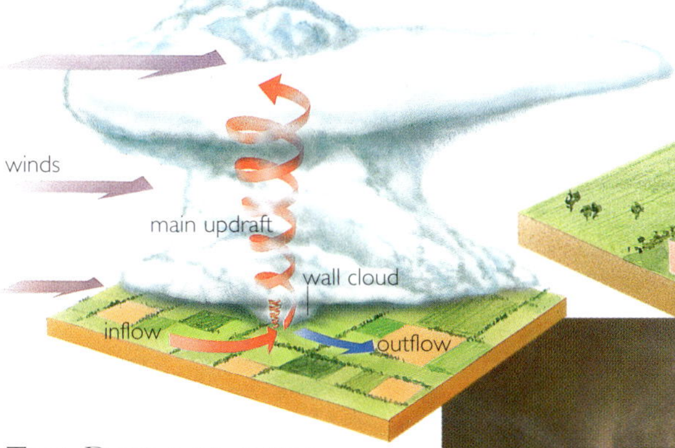

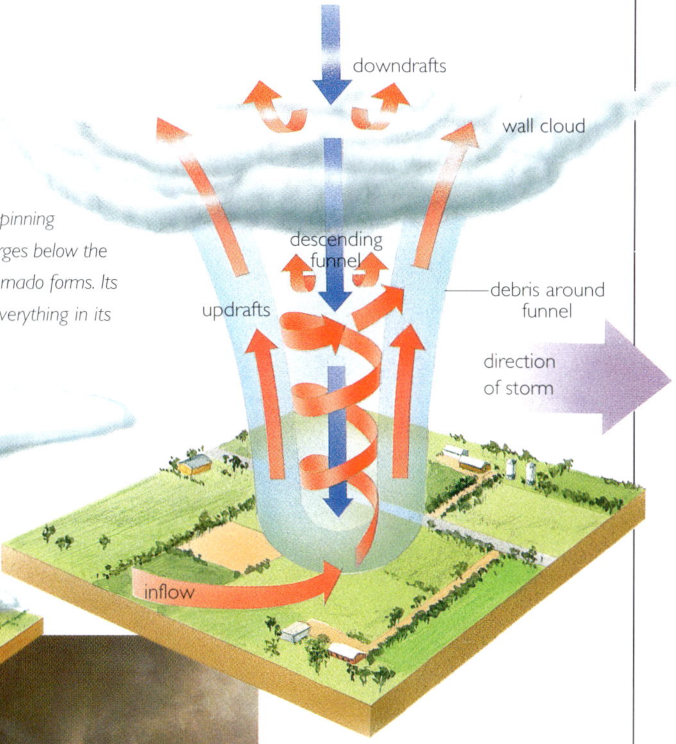

The Development of a Tornado

Tornadoes always develop from intense thunderstorms, such as supercell storms (see p. 49). The spinning motion usually begins when high-level winds blowing faster and in a different direction from low-level winds cause the whole storm system to rotate. Any rotating object moves faster as it is pulled toward its axis of rotation. So, as low pressure in the main updraft area of the storm draws in winds, they rotate more and more rapidly.

In some cases, the rotation is enhanced by a powerful, spinning, upward-moving column of air at the heart of the storm. This mesocyclone, as it is known, is caused by the interaction of warm and cold air currents in one part of the storm. Sometimes, the mesocyclone produces a protruding "wall cloud" on the underside of the storm. This is a clear sign of a developing tornado.

As the spin at the center of the storm intensifies, it begins to work its way down the main updraft toward the ground. To understand how this happens, imagine you are holding a rubber band taut and upright. If you twist the top of the band, the twist will slowly work its way downward.

Eventually, a column of rapidly rotating air emerges from the cloud base. This rotation may become visible as a funnel cloud if the pressure within it is low enough to cause condensation.

When the funnel hits the ground it becomes a fully fledged tornado. Its outline will be enhanced by the debris it sucks up, and the tornado may then take on various forms, from a thin, white cord to a thick, black mass.

Inside the Vortex

The tornado will move horizontally with its parent storm system at an average speed of 35 miles per hour (55 kph), although some tornadoes have been recorded traveling at up to 65 miles per hour (105 kph). It will be anywhere from 300 feet (90 m) to over ½ mile (800 m) wide and can cause a path of destruction from a few yards to hundreds of miles long.

Wind speeds inside a tornado are difficult to measure because instruments tend to be destroyed, but estimated top wind speeds are about 300 miles per hour (483 kph), while the updrafts may reach 180 miles per hour (290 kph).

The life span of a tornado can be from a few minutes up to an hour, but they usually last about 15 minutes. After the tornado reaches its maximum intensity, the funnel narrows and tilts toward the horizontal, and the path of damage becomes smaller. The funnel takes on a rope shape—called the rope stage—twists, and finally dies away.

Tropical Storms

Lasting from a few hours to as long as a month, the largest, most destructive storms known on Earth are born over the balmy waters of the tropics.

Tropical areas generate a great deal of heat, much of which is dispersed to middle latitudes by tropical storm systems. The most powerful of these are hurricanes (see p. 250).

The name hurricane is used in North America and the Caribbean. West of the International Date Line (180 degrees longitude), such storms are known as typhoons, while around the Indian Ocean and in Australia, they are called tropical cyclones.

How Storms Form

Tropical storms begin with large-scale evaporation and convection produced by very warm sea-surface temperatures. These processes give rise to large areas of cloud that often form clusters of thunderstorms. If the storms form far enough away from the equator for the Coriolis effect (see p. 31) to become significant (normally outside the band from 5 degrees south to 5 degrees north), they will begin to rotate, counterclockwise in the Northern Hemisphere and clockwise in the Southern Hemisphere.

A rotating storm system at these latitudes is the first sign of a hurricane. At this point, meteorologists can usually identify the system on satellite images and issue warnings.

The storm may then intensify as latent heat, instability (see p. 42), and warm ocean waters continue to fuel it. For it to become a hurricane, outward-spiraling, upper-level winds are needed to help take some of the rising air away from the storm and allow continued development.

If the system continues to move away from the equator, the spin will increase and

HURRICANE DISTRIBUTION *and the main storm paths are shown below. The season runs from June to November in the Northern Hemisphere and from November to May in the Southern Hemisphere. Signs indicating evacuation routes are posted in hurricane-prone areas of the United States, such as Florida (above).*

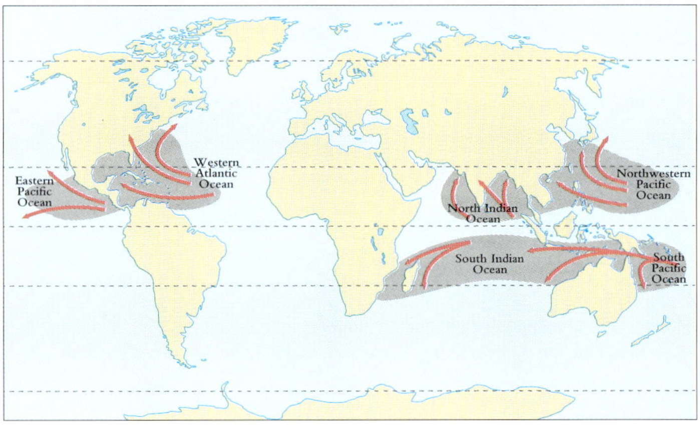

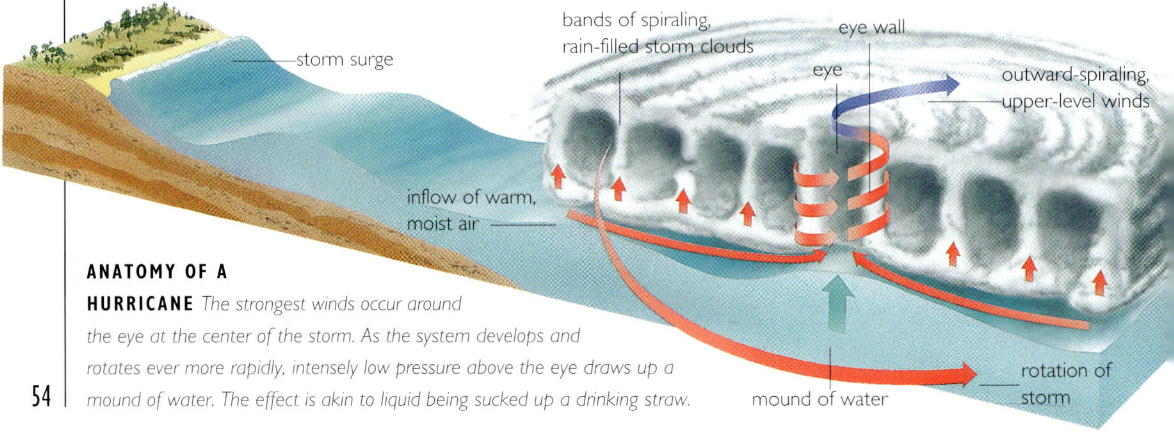

ANATOMY OF A HURRICANE *The strongest winds occur around the eye at the center of the storm. As the system develops and rotates ever more rapidly, intensely low pressure above the eye draws up a mound of water. The effect is akin to liquid being sucked up a drinking straw.*

winds will strengthen. When winds increase to 39 miles per hour (62 kph), the system is classified as a tropical storm. When winds reach 74 miles per hour (119 kph), the storm is then classified as a hurricane (typhoon or cyclone). Since 1979, weather services have given hurricanes alternate male and female names.

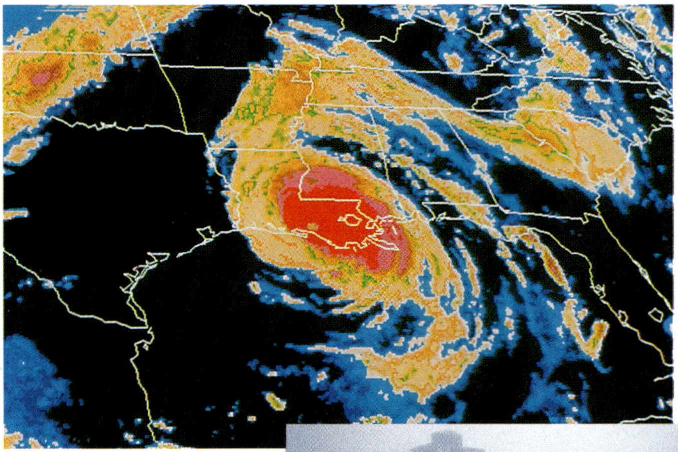

HURRICANE ANDREW *America's costliest hurricane struck Florida and Louisiana in August 1992 (above). The storm caused widespread devastation (right) and 23 deaths.*

The Eye of the Storm

The eye that forms at the center of a hurricane is a relatively calm area where the lowest air pressure occurs. The strongest thunderstorms and highest winds are usually found immediately outside this eye, in the eye wall—a ring of clouds that extends high above the sea. Hurricanes may rise many miles above the Earth, and can be hundreds of miles in diameter.

These destructive storms produce massive amounts of rain. Falls of 12 to 24 inches (300 to 600 mm) are not unheard of from a tropical system that reaches land. Wind gusts can exceed 150 miles per hour (242 kph), and the path of damage may extend from 200 to 500 miles (320 to 800 km) across. Tornadoes (see p. 52) can also form as the storm makes landfall.

The Storm Surge

The greatest danger to life from a landfalling hurricane is the storm surge. A mound of water forms under the center of a hurricane as the intensely low pressure draws water up. Over the ocean, this mound of water is barely noticeable, but it builds up as the storm approaches land. The surge's height as it reaches land depends upon the slope of the ocean floor at the coast. The more gradual the slope, the less volume of sea there is in which the surge can dissipate and the further inland the water is hurled.

Hurricanes weaken once they strike land because their supply of moist air is cut off, and friction between the storm and the ground slows the system. As they move away from the coast, hurricanes lose their tropical characteristics and become low-pressure systems, or "rain depressions". These often bring heavy rainfall to inland areas, causing widespread flooding.

LANDFALL *When a hurricane strikes land, property damage may be severe.*

THE SAFFIR–SIMPSON SCALE

Since the 1970s, the National Hurricane Center in the United States has used the Saffir–Simpson scale to classify hurricanes. The scale was developed by Herbert Saffir, an engineer, and Robert Simpson, a former director of the National Hurricane Center.

Category	*Pressure (hectopascals)*	*Wind Speed (mph [kph])*	*Storm Surge (ft [m])*	*Damage*
1	more than 980	74–95 (118–152)	4–5 (1.2–1.6)	Minimal
2	965–980	96–110 (153–176)	6–8 (1.7–2.5)	Moderate
3	945–964	111–130 (177–208)	9–12 (2.6–3.7)	Extensive
4	920–944	131–155 (209–248)	13–18 (3.8–5.4)	Extreme
5	less than 920	more than 155 (248)	more than 18 (5.4)	Catastrophic

Colors *in the* Sky

The everchanging colors of the sky are the result of the interaction between the water vapor and dust particles in the atmosphere and the colors that make up sunlight.

Color in the form of pigment does not exist in the atmosphere. Instead, the color we see in the sky results from the scattering, refraction, and diffraction of sunlight by particles in the atmosphere.

Color and Light

Sunlight travels through the solar system in straight, invisible waves. This "white" light is a mixture of all the colors—red, orange, yellow, green, blue, indigo, and violet—in the visible portion of the electromagnetic radiation spectrum. Each color of the visible spectrum travels at a different wavelength: red and orange have the longest wavelengths, while indigo and violet have the shortest.

When sunlight hits the atmosphere, the light waves are scattered in different directions by dust particles and air molecules. The shorter violet and blue waves are scattered more effectively than the longer orange and red ones. The effect is similar to what happens when ripples in water encounter a swimmer: small ripples are deflected while large waves continue past the obstacle undisturbed.

A mixture of violet, blue, green, and tiny amounts of the other colors is scattered across the sky. The combination of these colors is blue. The exact shade of blue will vary according to the amount of dust and water vapor in the air. Water droplets and dust particles enhance scattering, increasing the amount of green and yellow and turning the sky a paler blue.

CREPUSCULAR RAYS *The scattering effect of particles in the lower atmosphere renders these rays of sunlight visible.*

This is why the summer skies of densely populated European countries seem paler than those of vast, sparsely populated areas such as Australia and Africa.

Cloud Color

Clouds are white because all the colors of the spectrum are scattered by the water droplets that make up the cloud. The result of this is reconstituted white light. If the light does not pass through to the viewer, or another cloud is casting a shadow, clouds may appear gray or black.

BLUE SKIES *The atmosphere scatters the colors in sunlight one by one, beginning at the violet end of the spectrum. When the Sun is high in the sky (left), only the violet, indigo, blue, and a little green are scattered, producing a blue sky.*

SCATTERING OF LIGHT *creates both blue skies and white clouds (above). In the case of clouds, water droplets scatter all the colors of the spectrum equally (right). This results in reconstituted white light and hence white clouds.*

Changing Skies

As the Sun rises or sets, its light travels a longer path through the atmosphere. More of the colors at the red end of the spectrum are scattered, and the sky turns from yellow to orange to red.

The orange and red colors of sunsets can be intensified by air pollution, and ash and smoke from fires and volcanic eruptions (see pp. 266–71).

Scattering can create other effects. If an object in the lower atmosphere, such as a hill or a cloud, blocks out some sunlight, the remainder may appear as rays of light. These crepuscular rays, as they are known, are enhanced by the scattering of light in the air between the object and the viewer. The rays appear to diverge as they approach the observer; however, this is the same type of optical illusion that makes railway tracks appear to converge as they extend toward the horizon.

RED SUNSETS *When the Sun is low in the sky, its path through the atmosphere is longer, and yellow, orange, and red colors are scattered near the ground.*

Rainbows and Haloes

When light passes through a clear substance such as glass or water, it is bent at a slight angle, or refracted. Because colors of different wavelengths are refracted at different angles, this process causes the colors of the spectrum to separate.

A rainbow is produced when light is refracted by the edge of, and then reflected off the back of, raindrops. Each color emerges from the drops at a slightly different angle. Because the same color will emerge at the same angle from every one of the millions of drops in the cloud, the viewer sees distinct bands of colors.

A different combination of reflection and refraction creates optical effects known as haloes and sun dogs (see pp. 260–61). Another process known as diffraction, which involves the bending of light around the edge of an object, produces colorful rings known as coronas (see p. 258). These usually occur in middle-level clouds.

My heart leaps up when

I behold

A rainbow in the sky

"My Heart Leaps Up",
William Wordsworth
(1770–1850), English poet

Record Weather

The largest hailstone, the hottest place on Earth—records of extreme weather have been kept only for the past 180 years or so.

❶ *Highest measured wind gust: 231 mph (372 kph), recorded on 12 April 1934 at Mount Washington, New Hampshire, USA (at 6,288 feet [1,916 m]).*

The Radcliffe Observatory, Oxford, England (above left).

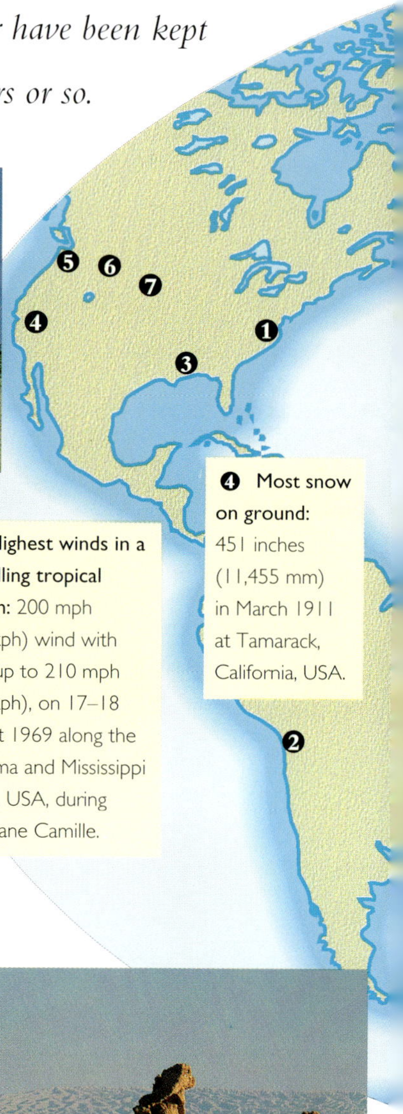

❸ **Highest winds in a landfalling tropical system:** 200 mph (322 kph) wind with gusts up to 210 mph (338 kph), on 17–18 August 1969 along the Alabama and Mississippi coasts, USA, during Hurricane Camille.

❹ **Most snow on ground:** 451 inches (11,455 mm) in March 1911 at Tamarack, California, USA.

Although sporadic records of significant weather have been kept over the centuries, systematic, routine weather observations did not begin until 1814, when the Radcliffe Observatory in Oxford, England, began recording changes in the weather. In the United States, a weather observatory founded by Abbott Lawrence Rotch on Great Blue Hill in Milton, Massachusetts, began to keep systematic daily records on 1 February 1885. The Blue Hill Observatory continues to keep weather records to this day, more than 110 years later—the longest continuously operational weather-observing station at the same location in the United States.

It's a Record

Weather extremes can be called records only if the weather station has an established climatology: that is, a long-term series of weather observations. The extremes recorded during the first year of readings should generally not be called records, unless they are compared with other local stations with long-established observations. But just how long weather stations should keep data before declaring records continues to be the subject of much debate: it has been suggested that a station should make weather observations for 10 years before calling an extreme reading a record.

World Records

Today, with worldwide stations keeping weather observations for many years, comparisons can be made to find the world records of extreme weather. Shown on these pages is a number of established world weather records.

❷ *Driest location: The Atacama Desert in Chile has virtually no rainfall (0.003 inches [0.08 mm] annually), except for a passing shower several times a century.*

5 Greatest measured annual snowfall: 1,224½ inches (31,102 mm), from 19 February 1971 to 8 February 1972 at Paradise, Mount Rainier, Washington, USA.

6 Greatest temperature change in one day: 100° F (55.6° C), a temperature drop from 44° F (6.7° C) to –56° F (–49° C), on 23–24 January 1916 in Browning, Montana, USA.

7 Most rapid temperature change: 49° F (27° C) in two minutes, a temperature rise from –4° F (–20° C) to 45° F (7° C), on 22 January 1943 in Spearfish, South Dakota, USA.

8 Greatest 24-hour rainfall: 73½ inches (1,869.9 mm), on 15–16 March 1952 at Chilaos, La Réunion, Indian Ocean.

9 Coldest annual average: –72° F (–57.8° C), at the Pole of Inaccessibility, Antarctica.

10 Hottest location: 136° F (57.8° C), on 13 September 1922 in Al' Aziziyah, Libya.

11 Greatest measured annual rainfall: 1,041¾ inches (26,461.7 mm), from 1 August 1860 to 31 July 1861 in Cherrapunji, Meghalaya, India (right).

12 Highest air pressure: 1083.5 hectopascals, on 31 December 1968 at Agata, Siberia, Russia.

13 Highest annual average rainfall: 467½ inches (11,874.5 mm), at Mawsynram, Meghalaya, India.

14 Largest hailstone: 2¼ pounds (1.02 kg), in a hailstorm on 14 April 1986 in Gopalganj district, Bangladesh.

15 Windiest location: winds reaching 200 mph (322 kph), at Commonwealth Bay, George V Coast, Antarctica.

16 Hottest annual average: 94° F (34.4° C), at Dallol, Ethiopia, 1960–66 (above).

17 Lowest air pressure: 870 hectopascals, on 12 October 1979 during Typhoon Tip; storm was 300 miles (483 km) west of Guam, Pacific Ocean (above left).

18 Coldest location: –128.6° F (–89.2° C) on 21 July 1983 at Vostok station, Antarctica (left).

The ocean of air in which we live and move, in which the bolt of heaven is forged, and the fructifying rain condensed, can never be to the zealous Naturalist a subject of tame and unfeeling contemplation.

LUKE HOWARD (1774–1864),
English pharmacist and meteorologist

CHAPTER THREE

WEATHER-WATCHING *through the* AGES

Early Civilizations

The first human societies were fascinated by weather phenomena and the movements of heavenly bodies, but they relied on mythology rather than rational analysis to explain them.

WEATHER GODS *Early northern European peoples regarded Thor (left) as the bringer of thunderstorms. Osiris (right) was the god of fertility in ancient Egyptian mythology.*

The first "meteorologists" were the shamans and priests of early communities, whose tasks involved appeasing the gods, who, it was believed, controlled all natural phenomena. The reputations of these mediators, and sometimes their very existence, depended upon their success in bringing good weather.

A Cradle of Civilization

Ancient Egypt was a perfect cradle for civilization. The weather was warm and sunny and the River Nile supplied water aplenty for irrigation. This meant, however, that the Egyptian communities, which were well established by 3500 BC, depended almost totally on the Nile for their prosperity. The Egyptians tried to use the movements of the stars as a guide to the annual rise and fall of the river as well as the extent of its periodic flooding.

This dependence on the Nile and the heavens found expression in two powerful gods—Ra (or Re) and Osiris. Egyptians believed that the Sun god Ra controlled the movements of heavenly bodies, traveling across the sky in his solar boat each day and returning through the underworld. Osiris was regarded as both the ruler of the dead and the source of fertility for the living—he controlled the sprouting of vegetation and the annual flooding of the Nile.

Divine Weather

Great rivers also supported the civilizations that emerged around 3500 BC on the flood plains of the lower Tigris and Euphrates rivers in Mesopotamia, and, later, in the Indus Valley in the Indian subcontinent. Although these cultures depended on rivers, their mythologies indicate that rain was also important. For example, the chief god of the Babylonians—whose kingdom in southern Mesopotamia flourished from about 2100 to 689 BC—was Marduk, originally god of thunderstorms but ultimately god of the atmosphere. In the Vedic religion of ancient India, Indra, god of the rain and storms, was among the most important deities.

In early northern European cultures, the Norse god Thor was thought to be all-powerful. His name comes from the Germanic word for thunder, and he was usually represented as a great warrior carrying a hammer, which symbolized a thunderbolt.

TONATIUH, *the Aztec Sun god, depicted here at the center of a calendar stone. In common with most early civilizations, the Aztecs saw the Sun as a divinity that controlled the movements of heavenly bodies, the weather, and, consequently, all human life.*

FLOOD WARNING *This seventh-century* BC *Assyrian tablet describes omens of heavy rains and flooding.*

... the windows of heaven were opened. And the rain was on the Earth forty days and forty nights.

Genesis 7: 11–12

Early Observations

Several early civilizations used astronomical observations to help them monitor seasonal changes in the weather. By 300 BC, Chinese astronomers had developed a calendar that divided the year into 24 "festivals" and specified the types of weather associated with each festival.

The rain gauge is probably the oldest meteorological instrument. The first reference to such an instrument appears in a work called *The Science of Politics* by Chanakaya, a minister of Chandragupta Maurya, who ruled India from 321 to 296 BC. Rainfall measurements were also made in Palestine 2,000 years ago.

Ancient Lore

Some peoples, particularly the Babylonians, tried to predict short-term changes in the weather. Their forecasts were based on astronomical observations, and the appearance of clouds and optical phenomena such as haloes. Among the forecasts recorded in the library of clay tablets of Assyrian King Ashurbanipal (c. 668–626 BC) is: "When a dark halo surrounds the Moon, the month will bring rain or will gather clouds ..." This was the beginning of the weather lore that would form the basis of forecasts for centuries to come.

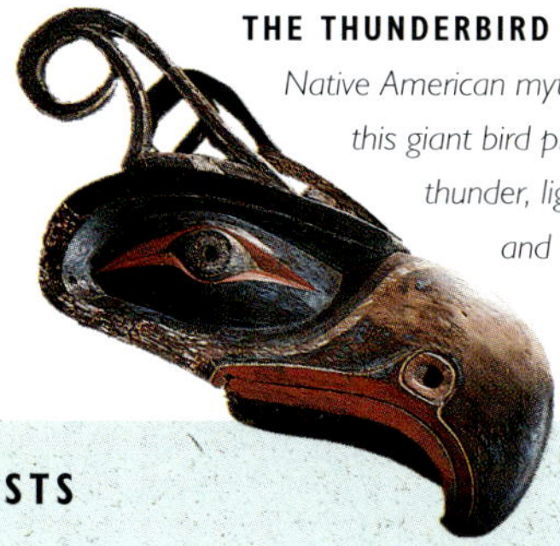

THE THUNDERBIRD *In Native American mythology, this giant bird produces thunder, lightning, and rain.*

BIBLICAL METEOROLOGISTS

Two famous "forecasters" receive prominent mention in the Old Testament—Joseph and Noah. According to the Book of Genesis, it was in Egypt that Joseph made one of the most famous long-range forecasts of all time. He interpreted the Pharaoh's dreams (which involved seven fat and seven thin cattle) and successfully predicted that seven good years would be followed by seven years of famine. Joseph's advice to store in times of plenty to provide for times of adversity is still relevant in drought-prone areas today.

In the biblical account of Noah and the Deluge, God gives Noah warning of the impending flood and instructs him to build an ark to save himself and his family. Thus, with a little help from divine instruction, Noah "predicts" the apocalyptic flood.

A similar account of a great flood occurs in Babylonian legend—the Gilgamesh Epic. It is likely that these stories were based on the same historical event. Excavations in Iraq have provided evidence of a great flood that took place on the flood plains of the Tigris and Euphrates rivers between 3000 and 2000 BC.

NOAH *loads the animals onto the ark, in this fresco from St Mark's Cathedral in Venice.*

Ancient Greece

The works of the great Greek philosophers, particularly Aristotle, established a lasting tradition of scientific inquiry and gave rise to the term meteorology.

Ancient Greece had no great rivers such as the Nile in Egypt, and Greek civilization was therefore dependent on rainfall for its water supply.

Ancient Greek mythology included numerous gods who were believed to personify and control all things celestial and terrestrial, including the weather. Ruler of the heavens was Zeus, who controlled the clouds, rain, and thunder. His brother Poseidon was the god of the sea and shores, and another brother, Hades (also called Pluto), ruled the underworld. Helios was the Sun god, and the winds were controlled by Aeolus.

To the Greeks, the appeal of this elaborate mythology seems to have been more aesthetic than religious. Their generally relaxed attitude to religion helps to explain the emergence of the Greek philosophers, who sought more rational explanations for natural phenomena.

The Early Philosophers

One of the foremost early philosophers was Thales of Miletus (c. 624–547 BC), who collected records from Babylonian astronomers and used them to successfully predict a solar eclipse in 585 BC. He considered that

water was the basis of all matter. Later, the philosopher and poet Empedocles (c. 495–435 BC) put forward the theory that all matter was composed of four elements—fire, air, water, and earth.

These and other early scholars made few significant discoveries in the physical sciences, but their work initiated a tradition of close investigation and rational analysis of all natural phenomena.

THE FOUR ELEMENTS OF EMPEDOCLES *(earth, air, fire, water), illustrated in a 1472 edition of Lucretius's* De rerum natura *(far left). The works of Theophrastus (left) remained a source of weather lore for 2,000 years.*

Aristotle and Theophrastus

The golden age of Greek scholarship reached a peak with Aristotle (384–322 BC), whose writings covered all aspects of human knowledge at that time. His treatise, *Meteorologica*, was an attempt to describe everything of a physical nature in the sky, air, sea, and earth, including all known weather phenomena. The title of this work gave rise to the term meteorology.

In *Meteorologica*, Aristotle made some remarkably accurate observations concerning wind and weather, and some shrewd deductions relating to natural phenomena, but he also made some significant errors of judgment—for instance: "The Earth is at rest, [wrong] and the

ZEUS, *hurling thunderbolts from his seat on Mount Olympus, depicted in a fresco by Giulio Romano (1499–1546).*

CAII PLYNII SECVNDI NATVRALIS HISTORIAE LIBER .I.
CAIVS PLYNIVS SECVNDVS NOVOCOMENSIS DOMITIANO SVO SALVTEM.
PRAEFATIO.

METEOROLOGICAL RELICS *The Tower of the Winds, which still stands in Athens today, dates from the first century* BC. *On each side of this octagonal building is a frieze depicting a male personification of the wind from that direction. Pliny's* Historia Naturalis *(right) gathered together writings on natural phenomena from Egypt, Babylon, Greece, and the Roman Empire.*

moisture about it is evaporated by the Sun's rays [right] ..."

When Aristotle retired, his pupil Theophrastus (c. 372–287 BC) continued his work. His book entitled *On Weather Signs* gives some 80 different signs of rain, 50 of storms, 45 of wind, and 24 of fair weather. Some were fairly reliable—for example: "Whenever there is fog, there is little or no rain." This is often correct because fog generally occurs with settled weather near the center of a high-pressure area. Others had no sound basis: "If shooting stars are frequent, they are a sign of either rain or wind; and the wind or rain will come from the quarter whence they proceed."

Theophrastus's attempts in other works to relate weather and clouds to wind direction were generally sound and based on solid observation.

Weather signs were further immortalized by the Greek poet Aratus (c. 315–245 BC) in his poem "Phaenomena". This contains a wealth of meteorological observations, and was regarded as a definitive work on weather in Greece and later in Rome.

THE LEGACY OF ARISTOTLE

The Romans too showed a strong interest in meteorology and were much influenced by Aristotle. The Roman writer Pliny the Elder (AD 23–79) compiled *Historia Naturalis*, a monumental encyclopedia that drew together the works of some 2,000 Roman and Greek authors, along with records and superstitions from Egypt and Babylon.

With the fall of the Roman Empire in the fifth century AD, the focus of civilization shifted to the Islamic world. Later, in the Middle Ages, Aristotle was rediscovered by European scholars (see p. 66). This proved to be a mixed blessing because scholars of the time were concerned more with interpreting Aristotle than with thinking for themselves.

HALCYON DAYS

The well-known expression halcyon days derives from a Greek myth relating to the weather. Two lovers, Ceyx and Alcyone, incurred the wrath of Zeus and Hera (king and queen of the gods). As a result, Ceyx's ship was wrecked in a storm, and he died. Alcyone, grief-stricken, then drowned herself in the sea. The dead lovers were subsequently transformed into kingfishers, birds believed to build floating nests. To protect their nests each year, Aeolus, Alcyone's father and guardian of the winds, suspended the winds for seven days before and after the winter solstice (midwinter's day, which falls on or about 22 December in the Northern Hemisphere). This calm, trouble-free period was subsequently known as the halcyon days.

THE MIDDLE AGES *and the* RENAISSANCE

The Renaissance saw a flourishing of scientific thought and ideas, and the invention of several meteorological instruments.

The progress of meteorology was stifled in the Middle Ages by an almost religious devotion to Aristotle and by the growth of astrometeorology. This began in Arabic countries, where seasonal weather predictions were based on the positions of stars and planets. These forecasts were included in almanacs—an Arabic term that originally meant the place where camels kneel but came to mean weather.

Astrometeorology gave rise to doomsday predictions, such as the "Toledo letter". In 1185, astronomer Johannes of Toledo, in Spain, predicted that in September of the following year all planets would be in conjunction, and storms, famine, and other disasters would result. Although this didn't happen, astrometeorology continued to flourish until the eighteenth century.

THE CONDENSATION HYGROMETER *(right) was invented by Ferdinand II. Ice was placed inside the device so that moisture would condense on the outside and drain into the measuring glass below. The amount of moisture collected indicated the level of humidity in the air.*

Almanacs were common in Europe and North America, reaching a peak of popularity in the seventeenth and eighteenth centuries. They often contained weather forecasts and tips for farmers as well as information on astronomical events and religious festivals.

THE RENAISSANCE

The Renaissance, which began at the start of the fifteenth century, saw numerous scientific discoveries that contributed to the development of meteorology. European explorers brought back a wealth of information on weather patterns over the oceans (especially in equatorial regions). Copernicus's (1473–1543) theory that the Earth rotates on its axis once each day and revolves around a stationary Sun once a year established the basis for a satisfactory explanation of the equinoxes, the solstices, and the seasons.

GALILEO, *author of groundbreaking works on astronomy, such as the* Dialogue Concerning the Two Great World Systems *(left), was also the inventor of the first thermometer. This became the model for the Florentine thermometer (above left) developed by the Accademia del Cimento, which, in turn, became a standard instrument throughout seventeenth-century Europe.*

Among contemporary scholars, the artist, engineer, and inventor Leonardo da Vinci (1452–1519) perhaps best personified the spirit of the Renaissance. Leonardo's journals contain numerous studies of weather phenomena and designs for meteorological instruments, including a hygrometer—a device for measuring humidity.

The mathematician and astronomer Galileo Galilei (1564–1642) was the first to develop a thermometer or, as he called it, a thermoscope. Unfortunately, Galileo did not leave any reports of observations made with this instrument.

Evangelista Torricelli (1608–47), a pupil of Galileo, made the first barometer. He filled a glass tube about 4 feet (1.2 m) long with mercury and inverted the open end into a dish of the same liquid. Torricelli noted that much of the mercury remained in the tube instead of going into the dish, and that the space above the mercury in the tube was a vacuum. He concluded that the mercury column was being supported by air pressure and that variations in the height of the column were caused by changes in air pressure.

French scientist and philosopher Blaise Pascal (1623–62) was one of the first to realize that these changes in atmospheric pressure might be related to changes in the weather, and this paved the way for the use of the barometer in weather forecasting. Pascal was also the first to demonstrate that air pressure decreases with altitude.

MEASURING HUMIDITY *This eighteenth-century hygrometer is based on principles described by Leonardo da Vinci (below) in his journals. The paper disks on the left absorb water vapor from the air. As the weight of the paper increases, the pointer rises, indicating the level of humidity on the scale on the right.*

The Accademia del Cimento

The Accademia del Cimento, or Academy of Experiments, was founded in Florence in 1657 by Grand Duke Ferdinand II of Tuscany (Ferdinando de' Medici) and his brother Leopoldo.

A number of improved instruments were developed under the auspices of the academy, including the condensation hygrometer and the Florentine thermometer.

In 1654, Ferdinand also set up the first meteorological observation network. Stations at Florence, Pisa, Parma, Curtigliano, Vallombrosa, Bologna, Milan, Innsbruck, Osnabrück, Paris, and Warsaw were equipped with standard meteorological instruments, which were used to record air pressure, wind direction, temperature, humidity, and current weather. These observations were sent to the academy for comparison. The network ceased when the academy closed in 1667.

FERDINAND II, *seated to the left of the table, presides over an experiment conducted at the Accademia del Cimento.*

THE AGE *of* REASON

During the late seventeenth and eighteenth centuries, the sciences of physics and meteorology advanced along a broad front.

The consistency and accuracy of meteorological observations improved steadily during the seventeenth and eighteenth centuries as a result of the development of new instruments and the growth of observation networks.

TEMPERATURE SCALES

The German physicist Gabriel Daniel Fahrenheit (1686–1736) spent most of his life designing and building weather instruments. He also developed the Fahrenheit temperature scale, still used in some countries, including the United States. This scale was calibrated on three points—the temperature of a mixture of water, ice, and common salt (0° F); the freezing point of water (32° F); and the temperature of the human body (assumed to be 96° F).

In 1742, Anders Celsius, a Swedish astronomer, introduced a scale that set zero at the boiling point of water and 100 degrees at the freezing point. This "upside-down" scale was designed to avoid negative temperatures in winter. In 1745, the scale was reversed by the Swedish scientist Carolus Linnaeus (1707–78). It is this scale that is known today as the Celsius scale.

LEARNED SOCIETIES, *such as the Académie des Sciences in Paris (above), became centers of meteorological research during the eighteenth century. Words such as fair and stormy first appeared on barometers (above left) in the late seventeenth century.*

MEASURING HUMIDITY

Humidity is more difficult to measure than temperature and early hygrometers were crude. In 1781, Horace Bénédict de Saussure (1740–99) discovered that human hair that had been boiled in a soda solution appeared to be a good indicator of humidity. Carefully treated hair is still used in hygrometers today.

An important breakthrough in hygrometry was made in 1802 by John Dalton (1766–1844), a British scientist. He demonstrated that the amount of water vapor required to saturate the air varies greatly with temperature. This led to the concepts of vapor pressure, saturation vapor pressure, and relative humidity (see p. 40).

NEW INSTRUMENTS

Atmospheric pressure is relatively easy to measure, and early barometers were fairly accurate. Most were based on Torricelli's barometer (see p. 67), but made use of a variety of liquids.

Robert Boyle (1627–91), an Anglo-Irish scientist, produced two models—

TRAIL OF DESTRUCTION *This early weather map shows the path of a storm that swept through northern France in 1788.*

a water barometer and a more portable type, the siphon barometer. The wheel barometer, invented by Robert Hooke (1635–1703), a colleague of Boyle's, used mercury and is believed to be the first to have had words—such as very dry, clear, variable, rain, and stormy—written on it. Hooke also invented a rain gauge.

Efficient pressure-tube wind gauges emerged in the 1740s. The best known of these, invented in 1891 by the British meteorologist W.H. Dines, is still in use.

THE EARLY WORKS *of Robert Boyle (left) were primarily concerned with the physical properties of air. Boyle was the first scientist to use the term barometer. Studies by English scientist John Dalton (above) provided the first explanation of condensation and proved that the aurora borealis is a magnetic phenomenon.*

INTERNATIONAL NETWORKS

In the eighteenth century, several meteorological networks were set up under the patronage of learned societies such as the Royal Society in Britain, the Académie des Sciences in France, and the Mannheim Meteorological Society in Germany.

A request in 1723 by James Jurin, secretary of the Royal Society, resulted in observations being received from England, continental Europe, North America, and India. In the 1730s, observation networks were set up in Siberia by scientists involved in Vitus Bering's Great Northern Expedition.

The most significant of the early networks was that established by the Mannheim Society. From 14 stations in 1781, the network gradually grew to include 39 observatories in Russia, Europe, Greenland, and North America. The society closed in 1799, but by then it had established procedures that were to prove invaluable when synoptic forecasting emerged in the nineteenth century (see p. 72).

THE FIRST AMERICAN METEOROLOGISTS

The first weather observations in the New World were those made by the Reverend John Campanius at the Swedes' Fort, near present-day Wilmington, Delaware, in 1644–5. The first thermometer and barometer measurements were made in the 1730s by John Lining, a Scottish-born resident of Charleston, in South Carolina.

The United States was fortunate that prominent figures such as Thomas Jefferson, Benjamin Franklin, and George Washington took a great interest in meteorology. Benjamin Franklin (1706–90) was a keen weather-watcher and the inventor of the lightning conductor (see p. 51). On the basis of newspaper reports, he demonstrated that in October 1743, a storm had moved northeastward from Georgia to Massachusetts. This is the first recorded analysis of the movement of a storm system.

Legend has it that Thomas Jefferson (1743–1826) acquired his first thermometer while writing the Declaration of Independence, and his first barometer, a few days after the document was signed. For over 50 years, Jefferson made regular weather records, and between 1776 and 1778 he and college professor James Madison made the first simultaneous weather observations in America.

George Washington also kept a regular weather diary. His notes for 13 December 1799 are thought to be the last words he ever wrote.

THOMAS JEFFERSON *(above) and James Madison made the first simultaneous weather observations in North America.*

Weather Lore

In the early nineteenth century, forecasting was still based mainly on weather lore, as it had been from Babylonian times.

The weather lore that had accumulated by the early nineteenth century was a curious mixture of sound common sense and sheer superstition, and included thousands of rules, quaint sayings, and proverbs. Some of this lore is still in use today.

EARLY LORE *was often based on relationships between weather and the winds. This woodcut by Albrecht Dürer (left) shows the global wind patterns as described by Ptolemy. The invention of the barometer (above) in the seventeenth century created a new category of sayings.*

Common-sense Lore

Much early weather folklore was based on the evident connections between winds, clouds, and the weather. For example, sheets of cirrus cloud often appear before a storm; the growth of convective cumulus clouds in the morning often leads to thunderstorms in the afternoon; rain is likely to follow if a halo is seen around the Moon—many such relationships are referred to in numerous observations, sayings, and proverbs.

Many of these sayings had originated with the Greeks, particularly Theophrastus (see p. 65), and had been nurtured and embellished through the Middle Ages. After the historic voyages of Columbus at the end of the fifteenth century, mariners extended this common-sense lore enormously to take account of the different wind systems and weather patterns they encountered around the globe.

It was discovered that many rules and proverbs were localized in their application, so that those that had originated in the middle latitudes of the Northern Hemisphere did not apply elsewhere. For example, the famous proverb "Red sky in the morning, sailor's warning; red sky at night, shepherd's delight" is based on the generally west-to-east movement of weather systems in the middle and high latitudes of both hemispheres, but does not apply in the tropics.

OLD FAVORITES *Among the most common weather proverbs are those relating to haloes around the moon (above) and red skies (right).*

The Moon and the Weather

Common among weather proverbs are those that link the Moon with cloud cover. These are reasonably good predictors. Frosty or foggy mornings often follow a cold, clear night, and it is on these nights that the Moon shines brightly—hence the saying, "Clear moon, frost soon". However, such proverbs tend to overlook the fact that frosty mornings often follow cold,

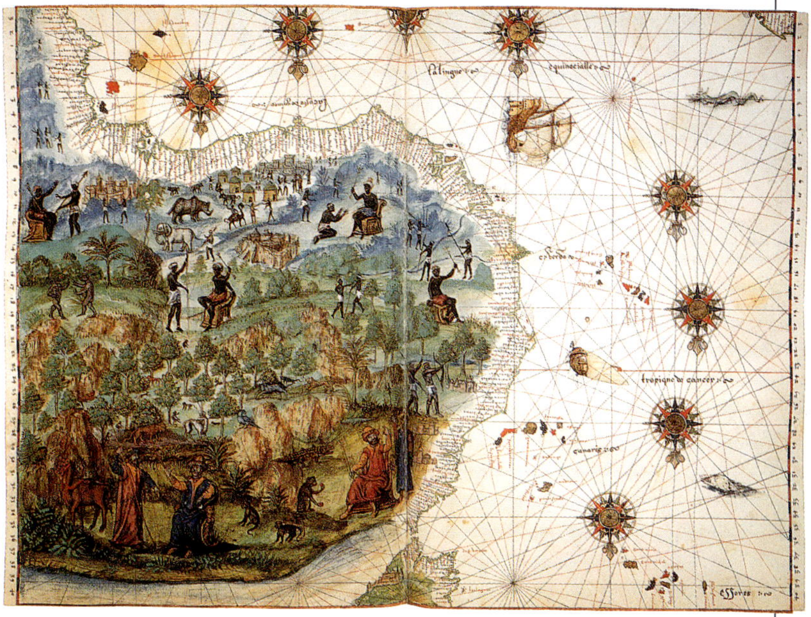

CHARTED WATERS *The voyages of explorers such as Christopher Columbus revealed the existence of weather patterns that were not accounted for in the lore of the mariners. Maps showing wind patterns thus became essential. In this 1547 map, West Africa is drawn "upside-down", as if seen from Europe.*

clear, moonless nights as well. Coronas and haloes (see pp. 258, 260) around the Moon indicate the presence of middle- or high-level clouds, which are often a sign of advancing rain or storms. Hence: "A circling ring of deep and murky red, soon from his cave the God of Storms will rise."

On the other hand, sayings that link the phases of the Moon with weather are superstition, not fact. Examples of these include references to the new Moon "lying on its back" or being "ill made" as a sign of wet weather.

Natural Signs

Sayings relating to the seemingly prophetic behavior of animals, insects, and plants abound in weather lore. For instance, cows are said to lie down before rain, and bees are supposed to return to the hive before a storm. However, most of these proverbs reflect the sensitivity of animals, insects, and plants to changes in current atmospheric conditions (most notably, changes in humidity) rather than any ability to predict future weather patterns (see p. 82).

Weather Lore and Forecasting

Through the centuries, mariners, farmers, and others attempted to make forecasts based on weather lore and personal observations. However, these predictions were often sadly astray. Inadequate communications meant that they did not know what was brewing over the horizon, and they were frequently caught unprepared by storms that swept in with little warning.

All this changed with the invention of the telegraph and the birth of synoptic forecasting in the 1860s (see p. 72).

SEASONAL PROVERBS

Generally, proverbs that link seasonal weather with subsequent good or bad seasons for crops are based on sound observations, but those linking the weather on particular days (or months) with subsequent seasonal weather have little or no statistical basis. Many of the following proverbs date back to the Middle Ages.

"January wet, no wine you get." (southern Europe)

"When March has April weather, April will have March weather." (France)

"St Swithin's Day [15 July], if ye no rain, for 40 days it will remain;
St Swithin's Day an ye be fair, for 40 days 'twill rain nae mair." (England)

"When it rains in August, it rains honey and wine." (France and Spain)

"When birds and badgers are fat in October, expect a cold winter." (United States)

"Frost on the shortest day [around 22 December] indicates a severe winter." (England)

A leaking May and a warm June, bring on the harvest very soon.

Scottish proverb

PIONEERS *of the* NINETEENTH CENTURY

The nineteenth century saw the emergence of synoptic forecasting, which would change the face of meteorology forever.

Synoptic forecasting involves the rapid collection and analysis of weather observations from as wide an area as possible. The concept of synoptic weather maps was developed by Heinrich Brandes (1777–1834) at the University of Breslau in present-day Poland. From 1816 to 1820, he prepared a series of maps based on observations made by the Mannheim Society network (see p. 69). These clearly showed high- and low-pressure systems over Europe. However, they were useless for forecasting, because by the time the information was gathered the meteorological situation had already changed.

It was only after the invention of the telegraph by Samuel Morse (1791–1872) in the 1830s that rapid communication, and hence synoptic forecasting, became practicable.

ROBERT FITZROY *(left) designed a barometer (right) for British sailors. Matthew Fontaine Maury (below).*

STORM WARNING NETWORKS

A meteorological network linked by telegraph was established in the United States in 1849 by Joseph Henry (1797–1878), secretary of the Smithsonian Institution. The data were collected by volunteers, and a daily synoptic map was displayed in the Smithsonian.

In Europe, during the Crimean War, a devastating storm in 1854 wiped out the Anglo-French fleet fighting the Battle of Balaclava. This made the allies aware that storm-warning services could be invaluable, and led to the creation of a meteorological network in France, which by 1857 was receiving data from all over Europe.

ROBERT FITZROY

In 1854, Admiral Robert FitzRoy (1805–65) was made head of the newly formed Meteorological Department of the United Kingdom Board of Trade. FitzRoy equipped British ships with a barometer that he had designed. He also began producing synoptic maps and, in 1861, started a storm-warning service for mariners. Initially, this was a great success and FitzRoy made his forecasts available to news-

DEVASTATING STORMS *during the Battle of Balaclava in 1854 encouraged England and France to set up meteorological observation networks.*

IMO *members gather for their 1879 congress (left). In the late nineteenth century, reports from meteorological stations throughout Britain were used to compile national records (below).*

papers. But as the inevitable errors occurred, criticism from the public and scientists became virulent. So depressed was FitzRoy that he committed suicide on 30 April 1865. Such criticism was to plague forecasters for the next century.

Matthew Maury

In the United States, Matthew Fontaine Maury (1806–73), head of the Navy's Depot of Charts and Instruments, used information from ships' log books to prepare charts showing the weather patterns over the oceans, as well as favorable sea routes. The use of Maury's charts resulted in significant reductions in the average times of sea voyages, and requests for his services increased enormously.

The IMO

Maury's influence helped him persuade the British and American governments to convene a conference in Brussels in 1853 to promote the international exchange of meteorological data. Other conferences followed and led to the founding of the International Meteorological Organization (IMO) in 1873—the year Maury died.

On 23 March 1950, the IMO became the WMO—the World Meteorological Organization (see p. 80).

EXPLORING THE UPPER AIR

Courageous balloonists risked their lives in order to study the skies and made a major contribution to meteorology by providing information on the winds and temperatures of the upper atmosphere.

Free balloon ascents began in 1783 and reached a peak with the intrepid Englishmen James Glaisher and Robert Coxwell. They made 28 flights over England between 1862 and 1866, during which they took hundreds of instrument readings. Their highest ascent, in September 1862, almost cost them their lives. As they rose past 29,500 feet (9 km), Glaisher passed out from lack of oxygen. At around 37,500 feet (11.25 km), a weakened Coxwell managed to force the balloon into a descent just before he too lost consciousness.

Soon afterward, the development of unmanned balloons carrying instruments aloft made these risky ventures unnecessary. The French meteorologist Teisserenc de Bort carried out hundreds of experiments with such balloons at his private observatory near Paris. These revealed something unexpected: the atmospheric temperature stops decreasing with height between about 29,000 and 42,500 feet (9 and 13 km) and, instead, starts to increase.

In 1902, de Bort managed to convince his colleagues that these figures were not errors, and a new atmospheric layer, the stratosphere (see p. 24), was discovered.

UP AND AWAY *The first manned balloon flights were made in Paris in 1783.*

THE AGE *of* SYNOPTIC FORECASTING

Throughout the late nineteenth and early twentieth centuries, the practice of synoptic forecasting was gradually refined.

Synoptic forecasting gathered momentum after about 1860 with the formation, around the world, of national meteorological organizations.

NATIONAL WEATHER SERVICES

In 1868 and 1869, storms on the Great Lakes in the United States sank or damaged more than 3,000 vessels and some 530 lives were lost. This led to a Resolution of Congress, which President Ulysses S. Grant signed into law in 1870, that established the nation's first official weather service, the meteorological division of the Signal Service—popularly known as the Weather Bureau.

VILHELM BJERKNES *coined the term fronts to describe the boundaries between warm and cold air masses.*

In addition to the United States, France, and the United Kingdom (see p. 72), national services were established by such countries as Austria, Denmark, Italy, Norway, Portugal, Russia, and Sweden. However, despite the best efforts of these fledgling organizations, the accuracy of weather forecasts improved very slowly.

THE BERGEN SCHOOL

A major step forward was made between 1918 and 1923 by a group of Scandinavian meteorologists led by Professor Vilhelm Bjerknes (1862–1951), known as the Bergen School. They put forward the theory that weather activity is concentrated in relatively narrow zones, which form the boundaries between warm and cold air masses. They called these zones "fronts", an analogy with the First World War battle fronts. It was subsequently confirmed that these fronts create much of our weather (see p. 34), and methods were developed that allowed meteorologists to predict their movements with considerable accuracy.

EARLY FRENCH WEATHER SERVICES *were particularly effective. By 1877, 1,230 French agricultural communities were receiving frost and storm warnings, enabling farmers to limit crop losses (left).*

WEATHER AND WAR

Both world wars forced governments to make greater efforts to monitor and predict weather patterns, as these could have a direct bearing on the outcome of battles. For example, the Allied D-Day landings, on 6 June 1944, proceeded during a temporary improvement in the weather that had been accurately forecast by American and British meteorologists.

The progress of meteorology was profoundly affected by the technology developed as part of the war effort.

Late-nineteenth-century experiments involving balloons carrying meteorological instruments (see p. 73) had led to the development in the 1930s of the radiosonde. This consists of a small instrument package attached to a balloon. The package includes pressure, temperature, and

WARTIME DISCOVERIES *B-29 bomber pilots (left) taking part in the first bombing mission to Japan in 1944 encountered powerful high-level winds, now known as jet streams (see p. 31). Radar systems (below), originally developed to track aircraft (below left), were subsequently used to monitor precipitation.*

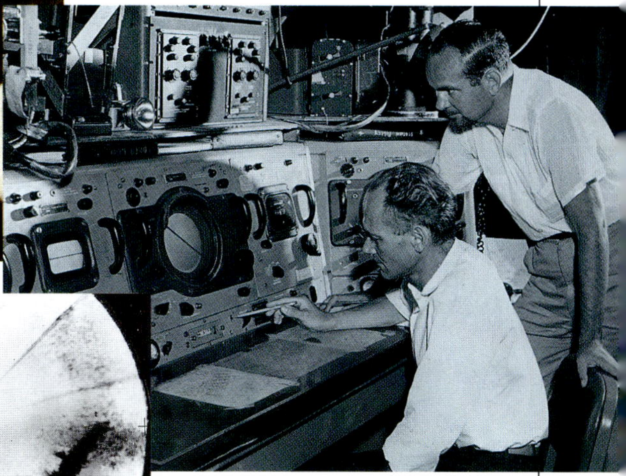

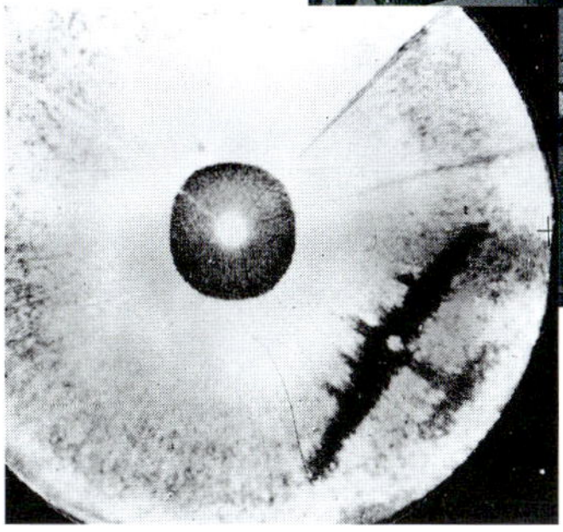

Nothing ... has proved so acceptable to the people, or has been so productive in so short a time of such important results, as the establishment of the Signal Service [Weather] Bureau.

CHARLES PATRICK DALY (1816–99),
American author and jurist

humidity sensors, and a transmitter that sends readings back to a ground station. During the Second World War, radiosonde networks expanded rapidly, with many flights extending well into the stratosphere. This provided scientists with a wealth of new information.

Balloons were also tracked from the ground with optical theodolites (an instrument that measures horizontal and vertical angles) in order to calculate wind speeds. However, the balloons often went out of range or were obscured by cloud. Radar (see p. 100), developed intensively during the Second World War to detect and track aircraft, provided the answer to this problem.

It was also found that radar could be used to track precipitation patterns. This helped weather stations to anticipate hurricanes, fronts, thunderstorms, and tornadoes.

UPPER-AIR CHARTS

Data from radiosonde and radar networks enabled meteorologists to construct weather charts for higher levels of the atmosphere. This was a highly significant breakthrough, because upper-level pressure, temperature, and wind velocity all have a strong influence on the weather patterns that we experience at ground level.

CARL GUSTAV ROSSBY

A disciple of Vilhelm Bjerknes, Carl Gustav Rossby (1898–1957) was one of the most influential meteorologists of the twentieth century. He was born in Sweden and went to the United States in 1926. He carried out pioneering work on the general circulation of the atmosphere and the meandering, long-wave patterns of westerly air flow in the upper troposphere that are now known as Rossby waves (see p. 31).

Rossby also worked on mathematical models for weather prediction, and provided the basis for the first successful numerical (computer) forecasting models. He is credited with predicting the existence of the jet streams and developing the main theories describing their behavior.

TOWARD *the* MODERN ERA

Modern technology, particularly electronic computers and meteorological satellites, has resulted in enormous improvements in the accuracy of weather forecasts and storm warnings.

LEWIS FRY RICHARDSON *(right) estimated that regular numerical weather predictions would require a factory of 64,000 mathematicians equipped with calculators. The logo of the World Meteorological Organization (above left).*

Numerical weather prediction is based on models in which the motion of the atmosphere and the physical processes taking place within it are represented in terms of mathematical equations. The idea of numerical forecasting was first formulated by Lewis Fry Richardson (1881–1953), a British mathematician, in his highly prescient 1922 paper, "Weather Prediction by Numerical Process".

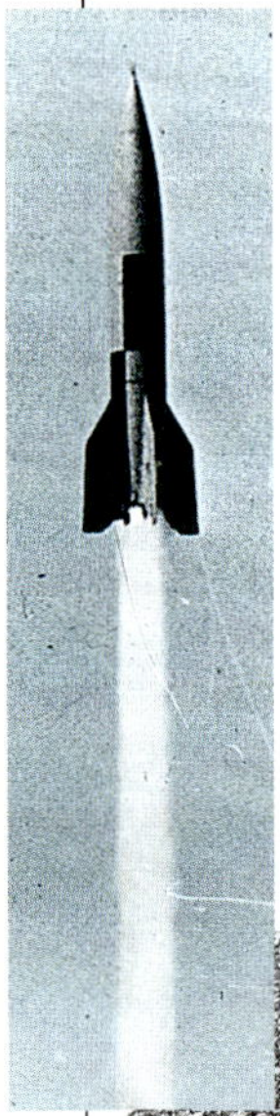

Richardson took many months to complete the arduous calculations necessary to produce a forecast for 24 hours ahead, and the pressure changes he predicted were 10 to 100 times too large, but he had made the first step toward accurate numerical weather forecasting.

MEETING THE CHALLENGE

Richardson's work highlighted certain fundamental obstacles to reliable numerical predictions: an enormous number of calculations had to be made very rapidly; the basic observational data were inadequate; the models were only crude representations of the atmosphere; and problems with the mathematical techniques could result in small errors being magnified as calculations proceeded.

Electronic computers have provided the answer to the calculation problem. In 1950 the first relatively successful numerical prediction was made in the United States. The Hungarian-born mathematician John von Neumann (1903–57) and his team produced this forecast using an early computer, the ENIAC (Electronic Numerical Integrator and Computer).

By April 1955, computer forecasts were being made in the United States on a regular basis. Initially, these were little, if any, better than those made by traditional means, but they improved rapidly as faster computers, more reliable data, and more sophisticated models became available.

METEOROLOGICAL SATELLITES

After the Second World War, United States scientists replaced the warheads of V2 rockets—rocket-propelled missiles invented by the Germans during the war—with cameras. They were dazzled by the results. At last, it was possible to view storms from space and obtain panoramic pictures of the weather.

Soon afterward, on 1 April 1960, the polar-orbiting satel-

V2 ROCKETS *gave the first views from space of the Earth's weather systems.*

lite TIROS 1 (Television Infrared Observation Satellite) was launched. It took 23,000 pictures of the Earth and its clouds over a period of 78 days. Meteorologists the world over were ecstatic.

Soon, the value of satellites was made clear to the public too. In September 1961, satellite images of Cyclone Carla led to the evacuation of more than 350,000 people from their homes along the Gulf coast—the largest mass evacuation in the United States to that time.

Throughout the 1960s satellite meteorology advanced at an astonishing pace. Improved infrared technology allowed pictures to be taken at night. By 1963, photographs could be obtained direct from satellites as they passed overhead, and in 1966 the first of the geostationary satellites—which hover over the equator and can see about one-third of the globe—was launched.

TIROS 1, *the first polar-orbiting satellite, launched on 1 April 1960.*

The World Weather Watch

The success of satellite technology inspired nations to cooperate in the collection of data. In 1961, President John F. Kennedy invited other countries to join the United States in developing an international weather-prediction program. Although this was at the height of the Cold War, 150 countries, including the USSR, responded positively, and the World Weather Watch (WWW) was set up under the auspices of the World Meteorological Organization (WMO) in 1963. Under the scheme, members of the WMO exchanged data from their observation networks on a regular basis, thus facilitating the preparation of global weather charts.

The World According to GARP

Members of the WMO also participated in the Global Atmospheric Research Programme (GARP). This involved a number of major meteorological surveys, including the Global Weather Experiment, the largest international scientific field exercise ever undertaken. For one year, from 1 December 1978 to 1 December 1979, the technological resources of WMO members were deployed continuously in order to subject the atmosphere to the most thorough analysis possible. The resulting data clarified our understanding of global weather patterns and allowed scientists to refine existing numerical models.

THE ENIAC COMPUTER, *used to produce the first successful numerical forecast, was a dinosaur: it consisted of 18,000 vacuum tubes, 70,000 resistors, 10,000 capacitors, and 6,000 switches. The machine's power source occupied about half as much space as the computer itself.*

... weather-lovers are part scientist, part poet. They rejoice in the forms and colors that glorify the weather. They delight in extremes.

Knowing the Weather,
T. MORRIS LONGSTRETH (1886–1975), American author

CHAPTER FOUR

FORECASTING TODAY

Weather Organizations

National and international organizations observe and predict the weather, working together in a spirit of global cooperation.

National weather services operate in almost every country, forming the hub of a complex web of weather-related activities that are central to our lives. The services are designed to respond to the particular climate of their country and so concentrate on whichever aspects of the weather have most impact on the running of the country: for example, hurricane forecasting in North America, or rainfall forecasting in Australia. Supplemented by various levels of government assistance, the services draw on the combined efforts of universities, industry, and their own research facilities.

The personnel involved in monitoring and forecasting the weather have a wide range of skills. They design, operate, and interpret the output of a variety of instruments, and then translate the information into forecasts.

COMPUTERS, *which can process huge amounts of data, are now a key forecasting tool at national services such as the Japan Meteorological Agency (above).*

MEASUREMENTS *are taken from a wide range of locations. In Australia, for instance, there are weather stations in the busy streets of Melbourne, Victoria (right) and the remote outback of Western Australia (below). The logo of the UK's Royal Meteorological Society (top left).*

Powerful computer systems are used to perform scientific and mathematical analyses of the data. A massive amount of information must then be reduced to an accessible and useful form.

Sophisticated links with the print and electronic media provide quick, efficient ways of delivering information to the general public. As well as standard forecasts, this information includes warnings of hazardous weather such as heavy rain or floods, fog, ice, snowfall, or high winds.

National weather services also provide specialized information for various sectors of the community, including farmers, the armed forces, and public transport authorities.

Global Weather

As weather is global in nature, worldwide cooperation is essential. Most national weather services concentrate on producing detailed forecasts for their own countries one to two days ahead, but they all contribute to international work.

The national weather services that are particularly involved in global forecasting are the American National Weather Center, located at Camp Springs, Maryland; the British Meteorological Office, at Bracknell, England; and the European Centre for Medium-Range Weather Forecasts (ECMWF), near Reading, England. The ECMWF is required to produce weather

NATIONAL WEATHER SERVICES *concentrate on forecasting the conditions that affect their region, such as snowstorms in North America (left) and monsoon rains in the subtropics (below).*

forecasts 1 to 10 days ahead for its 17 member countries. Because it does not have to produce short-range forecasts, it is able to spend time collecting and assimilating data, and is regarded as a leader in medium-range forecasting. It shares its results with weather services around the world.

The World Meteorological Organization (WMO) is a specialized agency of the United Nations. It has over 170 member countries and runs three World Meteorological Centers: one in Melbourne, one in Moscow, and one in Washington DC. Established in 1951, its main tasks are to improve weather observations throughout the world, provide a more uniform weather-reporting system, and implement a better distribution of weather information. At a practical level, the WMO plays a central role in establishing basic standards for meteorological procedures, such as reaching agreement on codes for transmitting data. It also has eight technical commissions, which act in parallel with national weather services. These commissions cover synoptic meteorology, aviation meteorology, climatology, physics of the atmosphere, agricultural meteorology, hydrometeorology, maritime meteorology, and instruments and methods of observation.

World Weather Watch (WWW) is a global meteorological system that uses the facilities and operations of the national weather services of WMO members. WWW collects data from five stations on Earth, a series of polar-orbiting weather satellites, and specialized networks such as radar stations. It also receives observations from some 12,000 land stations, over 7,000 ships and oil rigs, 700 stations making upper-air observations with radiosondes (see p. 100), and many commercial airliners. All this data is transmitted in code via teletype and radio links to regional or national centers, and is then fed into the high-speed Global Telecommunications System connecting the three World Meteorological Centers of the WMO. The three centers assemble the data to create global synoptic maps and also enter it into computer models to produce global forecasts. The maps and forecasts are distributed every six hours to national weather services.

NOAA, *an agency of the US Commerce Department, gathers data about our oceans and atmosphere, and runs the National Weather Service. The NOAA logo (right) and a meteorologist (below).*

PREPARING *a* WEATHER FORECAST

An extensive network of people and equipment provides the data needed to produce accurate forecasts.

TRACKING THE WEATHER *A hand-drawn map being prepared (top left). Reading a thermometer on the weather-reporting ship* Aurora Australis *(above). Keeping track of radar signals (right).*

Meteorologists distinguish between skill and no-skill forecasting methods. If we consider rainfall prediction, two basic no-skill methods appear to give impressive results. The first is the persistence method, which is simply forecasting tomorrow's rain to be the same as today's. In middle latitudes, this typically gives results of about 70 percent accuracy, but of course fails to predict changes.

The other no-skill method, the climatological method, uses long-term averages. If, for example, the statistics for a particular location show that during January there is an average of 10 rainy days, then we would forecast rain every third day. Our forecasting accuracy would again be about 70 percent for many middle-latitude locations.

These methods take no account of the actual weather. For a forecasting technique to demonstrate skill, it must be more accurate than these no-skill approaches.

Skill methods make use of data from a wide variety of sources to put together a comprehensive picture of the current state of the atmosphere. The information is entered into global computer models (GCMs), which are then used to simulate future weather patterns. Both the current and predicted situations are represented on weather maps (see p. 84).

THE HUMAN ELEMENT

Human observers are at the core of the weather-measuring system, providing forecasters with raw data. The worldwide, land-based network involves weather-service professionals, people working in weather-sensitive industries, and private citizens who have become official observers.

MEASURING WIND SPEED *near France's Dumont d'Urville weather station in Antarctica.*

Every cloud engenders

not a storm.

King Henry VI Part III,
WILLIAM SHAKESPEARE
(1564–1616),
English playwright

FROM SPACE *Our planet as seen from a geostationary satellite, which remains over the same spot on the Earth (left). A storm in the Bering Sea (below), as seen from a polar-orbiting satellite.*

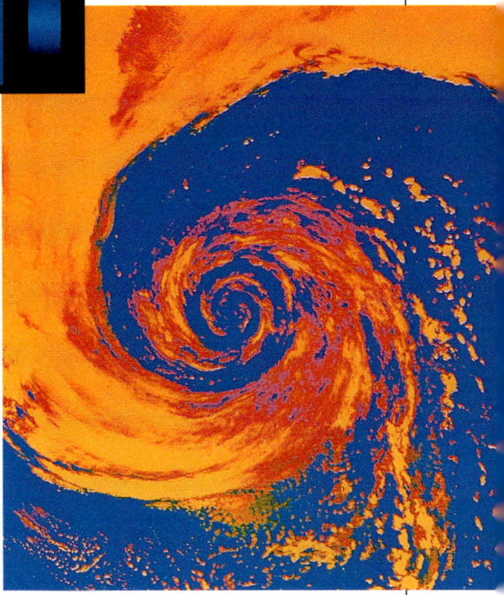

Observers follow a well-established routine. Readings are taken at standard times—usually every six hours, but, in some cases, every three hours. Various measurements may be made, but the essentials are: humidity; maximum and minimum temperatures; rainfall; wind speed and direction; direction of cloud movement, type and height of clouds, and how much sky is covered; and air pressure (adjusted to sea level).

As quickly as possible, all this information is converted to a numerical code and transmitted, usually by high-speed teletype, to a regional center. Here, all the reports from a specific locality are collated and forwarded to the central forecasting agency.

Measurements are also provided by crew on board ships, oil rigs, and airplanes. Furthermore, electronic instruments supply a wealth of information (see p. 100). Automatic weather stations can be programmed to supply regular readings. Upper-atmosphere measurements are made by radiosondes—packs of electronic instruments carried by balloons. Rainfall information is provided by radar, and satellites measure infrared radiation (heat) rising from the Earth's surface.

PRESENTING A TRUE PICTURE

Forecasters must start their computations with as precise a picture of the global weather as possible. The larger the errors when they begin, the more rapidly their forecasts cease to be of use (see p. 104).

However, many more measurements are made on land than at sea or in aircraft. Satellites help to balance the input of data, but they also have their limitations. For example, measurements for particular places can only be made when orbiting satellites are directly overhead.

The amount of data that must be processed each day is mind-boggling. One satellite alone can provide 150,000 observations a day. Within the accepted six-hourly or three-hourly reporting interval, any given forecasting agency has to collect material from all its various sources, check it for errors, reduce it to a standard format, and then feed it into GCMs. In practice, about 90 percent of data is processed within three to four hours.

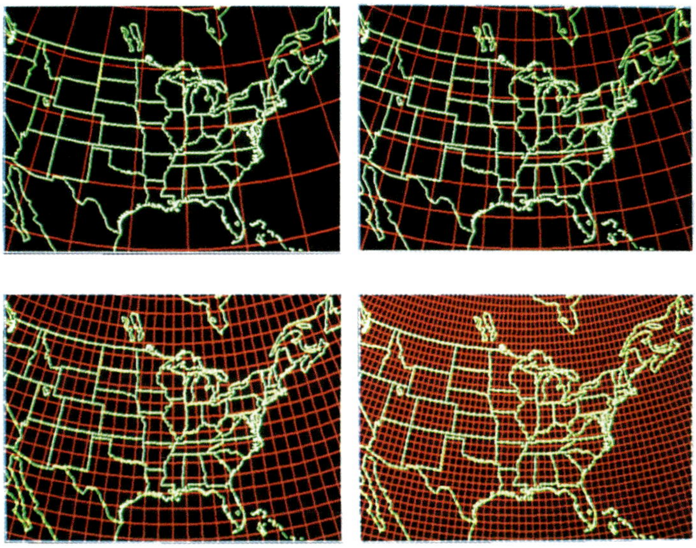

GLOBAL GRIDS *(left) mark data-entry points for computer forecasting models. The resolution of the grid varies according to the purpose of the model, ranging from an eight-degree grid for general climate models to a one-degree grid for more detailed forecasts.*

WEATHER MAPS, *also known as synoptic charts, are important forecasting tools. They may be drawn by hand (right) or prepared by computer. Weather forecasters at the Adelaide Meteorological Office, South Australia (below right).*

Drawing the Map

Until recently, weather maps were drawn by hand, but this task is now largely performed by computers. In either case, the basic procedure is the same. The weather conditions in each location, as measured by observers and automatic equipment, are plotted on a surface map. Standard symbols (see selection, facing page and below) are used throughout the world so that all users can understand the data.

Lines are drawn between points of equal air pressure to form isobars. Closed isobar curves indicate areas of high and low pressure, which are labeled (usually with the letters H and L). Next, cold and warm fronts—the boundaries between air masses—are identified with the assistance of satellite images and marked on the map. Areas of rainfall may be shaded. (See p. 88.)

Weather maps are also produced for the upper levels of the atmosphere, using a wide range of information supplied by radiosondes, aircraft, radar, and satellites.

Interpreting GCMs

The largest supercomputers running GCMs can now perform more than a billion calculations a second. Within an hour they can produce a set of forecasts to meet the requirements of a variety of users ranging from the media to the aviation industry.

Human interpretation of the output of GCMs is an essential part of producing forecasts. Using features of satellite and radar images that cannot be incorporated in the models, it is often possible to recognize where the simulations are going wrong. Forecasters will also have satellite and radar data that arrived after the deadline for the computer analysis. Their interpretation of this material can provide refinements, which are particularly valuable in predicting weather a day or so ahead.

Judgments of this nature must be exercised as part of a hectic schedule. The final forecasts are issued just as the next round of measurements is starting to arrive and the whole process is beginning all over again.

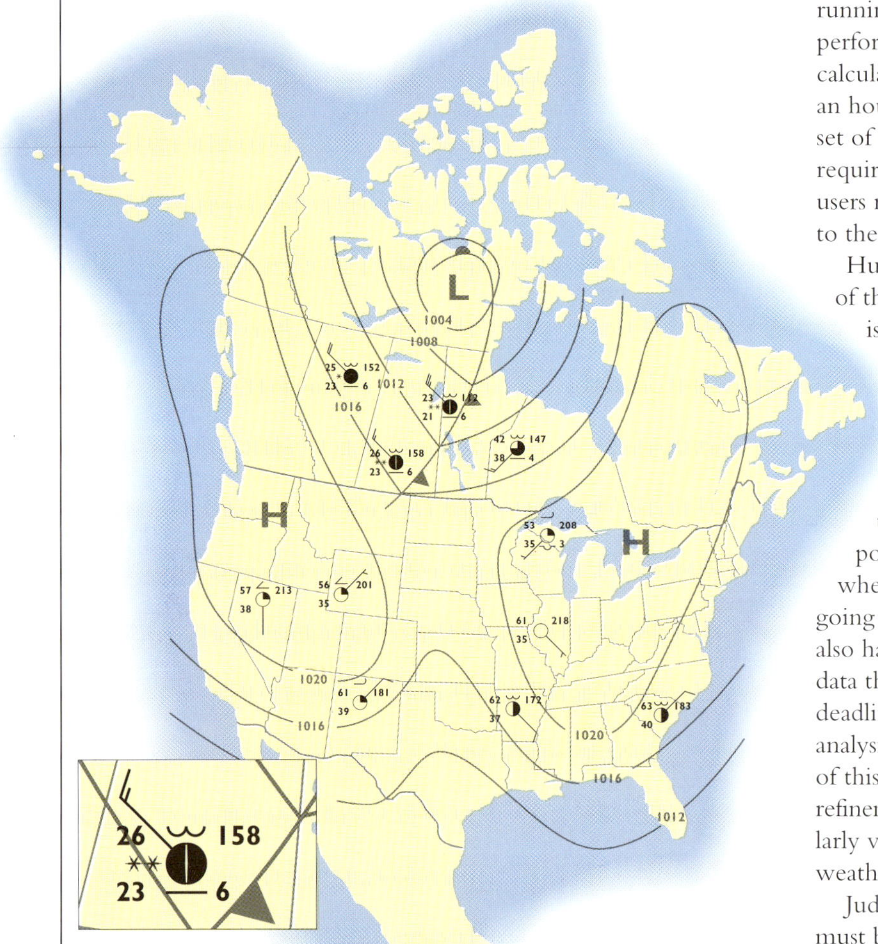

THIS WEATHER MAP *(above) is a simplified version that shows how symbols are plotted to create an overall picture. The symbols in the inset represent:*

- ◐ *sky nine-tenths covered*
- *wind from NE at 15–20 mph*
- ✱✱ *steady, light snow*
- ᴗ *altocumulus in sky*
- **—6** *stratus covers seven-tenths of sky*
- **158** *air pressure is 1015.8 hpa*
- **23** *dewpoint is 23° F*
- **26** *temperature is 26° F*

INTERNATIONAL WEATHER SYMBOLS

Current weather

light drizzle
steady, light drizzle
intermittent, moderate drizzle
steady, moderate drizzle
intermittent, heavy drizzle
steady, heavy drizzle
light rain
steady, light rain
intermittent, moderate rain
steady, moderate rain
intermittent, heavy rain
steady, heavy rain
light snow
steady, light snow
intermittent, moderate snow
steady, moderate snow
intermittent, heavy snow
steady, heavy snow
hail
freezing rain
smoke
tornado
dust devils
dust storms
fog
thunderstorm
lightning
hurricane

Sky coverage

no clouds
one-tenth covered
two- to three-tenths covered
four-tenths covered
half covered
six-tenths covered
seven- to eight-tenths covered
nine-tenths covered
completely overcast
sky obscured

Low clouds

stratus
stratocumulus
cumulus
cumulus congestus
cumulonimbus calvus
cumulonimbus with anvil

Middle clouds

altostratus
altocumulus
altocumulus castellanus

High clouds

cirrus
cirrostratus
cirrocumulus

Wind speed

mph	*kph*
calm	calm
1–2	1–3
3–8	4–13
9–14	14–23
15–20	24–33
21–25	34–40
55–60	89–97
119–123	192–198

Disseminating Weather Information

Every day, millions of observations and calculations are condensed into forecasts of just a few words and images.

Meteorologists present their forecasts in a variety of ways. Simplified forecasts are distributed to the public through the media. These provide general information, such as global and national predictions, and also give more detailed analysis of local conditions. Some groups and industries require information about particular weather conditions. This is supplied through a range of services.

In the Media

Television, radio, and newspapers provide us with considerable amounts of information about the weather. In newspapers and on radio, weather forecasts are confined to the bare essentials, although most papers include a weather map and many give a recent satellite image. Television forecasts do not usually offer much more information, but they may visually overload the viewer by providing several weather maps for the days ahead, together with moving three-dimensional satellite images and radar pictures of rainfall distribution.

Special Services

Media information can be supplemented by data from a variety of additional sources, including fax, telephone, and computer information services such as those on

the Internet. These services provide detailed forecasts for specific areas—for example, mountainous regions or inshore waters. They also cater for certain interest groups, such as mountaineers and sailors, and industries, such as retail stores (people tend not to go shopping during bad weather) and electricity-generating companies.

Anyone with a keen interest in the weather can gain access to a wealth of up-to-date information. In the United States, the National Oceanic and Atmospheric Administration offers continuous weather information direct from about 350 National Weather Service offices, and updates the information every one to three hours.

People who are unable to obtain published weather maps, such as sailors, can use their radios to pick up shipping bulletins issued by national weather services around the world. In many countries, more comprehensive

FORECASTS *appear in various forms in newspapers throughou the world (top). The Weather Channel in the USA broadcasts forecasts 24 hours a day (above). Weather presenters, such as Willard Scott of NBC (right), are a familiar sight on our screens.*

EARLY WARNING *of severe weather can help save lives and sometimes gives people enough time to safeguard property. Meteorological services around the world provide valuable warning of conditions that may cause flooding (right) and wildfires (below).*

information is broadcast in Morse code on radio and also in coded form via radio teleprinters. Details of these services, and the frequencies on which they operate, are available from your national weather service or from the World Meteorological Organization (WMO).

Weather offices supply meteorological information directly to some industries. The aviation industry depends on detailed and accurate forecasts to ensure safety and economic viability. For instance, if a forecaster predicts that fog is likely to form at a particular airport, then all incoming traffic will be required to carry enough fuel to reach the nearest alternative airport.

Severe Weather Warnings

A striking feature of improved forecasting is the growing ability of weather services to issue warnings in good time. Computer models can now anticipate the explosive development of major storms, such as the Great East Coast Blizzard of March 1993, long before there is any significant feature on the current charts, and predict with increasing accuracy the course of hurricanes. These improvements have bolstered the confidence of forecasters and encouraged the public to heed weather warnings. Forecasting severe storms and tornadoes remains problematic, as they are often too small to be picked up by computer models (see p. 104).

Using the Forecasts

As you learn more about the weather, you can exercise some of the judgment that the forecasters use and become skilled at interpreting the meteorological data presented by the media (see p. 88). You can then use local knowledge to fill out the picture, which will help when planning outdoor activities from lighting a barbecue in the garden to hiking in the backcountry.

When severe weather threatens it is vital to stay in touch with forecasts. If you combine the information they provide with what you are experiencing—strengthening winds, lowering clouds, even funnel clouds on the horizon, or snow or heavy rain starting to fall—then you can respond effectively to warnings.

However confident you may be about your weather-watching abilities, never disregard warnings given by professionals of dangerous conditions, especially as these can sometimes develop with frightening rapidity.

HURRICANE *warning flags flying in Florida, USA (right). Tornadoes, such as the one that caused this wreckage in Saragosa, Texas, USA (far right), tend to be too small to be picked up by computer models.*

Reading Weather Maps

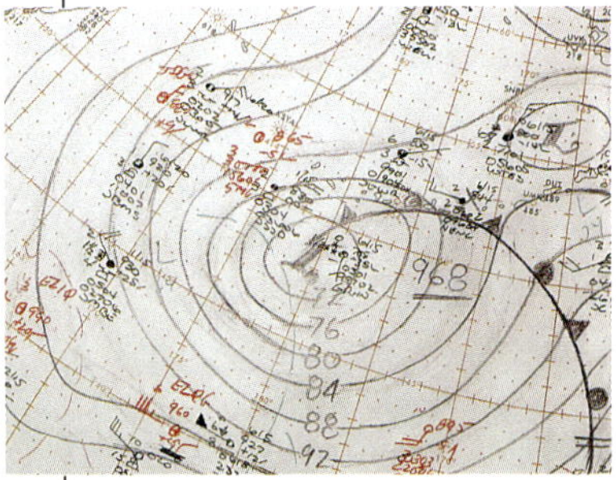

Once you understand the basic elements of a weather map you will be better able to interpret forecasts.

HIGHS AND LOWS *This hand-drawn map (top left) shows a low. Air pollution levels in cities (above) often increase during still conditions caused by highs.*

Weather maps are the result of condensing a huge amount of information into a standard format. They record the most recent observations or the output of forecasting models, and show either the current situation (synoptic chart) or the predicted situation at some time ahead (prognostic chart).

What's on the Map

The first features to look for on a weather map are the areas of high and low air pressure. These are usually labelled with an H or an L at the center of the system.

Many weather maps also show isobars, which are lines of equal air pressure. On some maps, each isobar will be labelled with a number representing the air pressure in hectopascals (see p. 96).

In a high-pressure system, the air pressure increases toward the center, whereas in a low, it decreases toward the center.

Close spacing of isobars indicates strong winds, which are normally associated with low-pressure systems. Conversely, where the bars are far apart, relative calm prevails, usually associated with high-pressure systems (which often also produce clear skies). In winter, highs can lead to low temperatures at night (see p. 35) and, in both winter and summer, they often create stagnant conditions that cause pollution levels to soar.

It helps to understand how winds are flowing around the main highs and lows on a map and whether they are drawing in air from lower or higher latitudes. You can tell this by looking at the isobars. In the Northern Hemisphere, winds will flow in a counter-clockwise direction around lows and in a clockwise direction around highs. (Note that these directions are the opposite in the Southern Hemisphere.) By determining the wind direction and following the isobars out from the center of the highs and lows, you should be able to tell where

NEWSPAPERS *(right) usually contain simplified maps, which may or may not show isobars. Most will represent the edge of a cold front (above) as a line with triangles.*

WEATHER MAPS AND SATELLITE IMAGES

Combining information available in weather maps and satellite images can produce additional insights into what the weather is doing. Both regularly appear in the media, and are now also available via fax services or the Internet. The satellite images clearly show cloud formations, while the pressure patterns in weather maps depict how the atmosphere is moving, where the major fronts are, and where the air masses are coming from. Together, they show how the atmosphere generates weather.

The example shown here is the "Superstorm" of March 1993, which moved along the east coast of the United States. Beginning as a "normal" low-pressure cell over the Gulf of Mexico on Friday 12 March, the system then deepened dramatically and moved northeast over Florida in the following 24-hour period.

In the satellite image (top right), we see cold air being drawn down from the far north to the southern states, and a well-developed squall line (see p. 48) preceding the frontal system. The weather map (middle right) shows the storm centered over southeastern Georgia with the cold front (the line with the triangles) and the squall line (the dotted line) just to the east of Florida.

The Superstorm produced flooding and tornadoes over Florida, and unprecedented snowfalls over many eastern states (bottom right), including 13 inches (32.5 cm) at Birmingham, Alabama.

The National Weather Service had access to regularly updated weather maps and satellite images as well as advanced radar systems and computer simulations, and was able to issue comprehensive warnings almost 24 hours before the first snow fell.

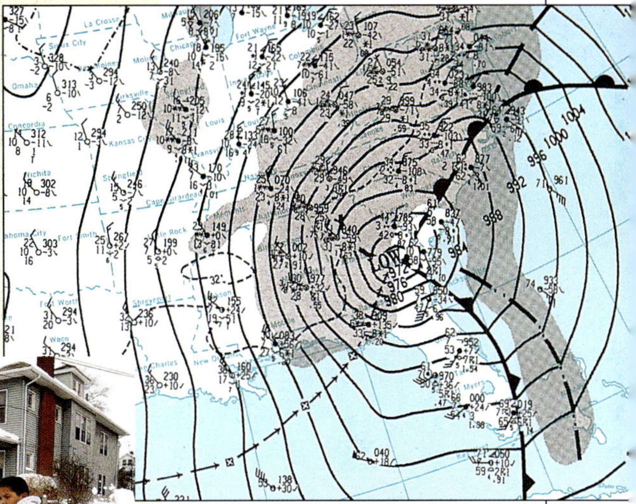

the winds in your part of the world are coming from.

The source of a wind will have a major influence on the type of weather it brings. For example, air from high latitudes in winter is cold and dry, but if it passes over oceans it is warmed and will become cloudy. This is why the coldest spells in North America and Europe occur when arctic air plunges down from northwest Canada or out of Siberia. Conversely, air drawn from low latitudes will be warm and humid (see p. 30).

Weather maps also show fronts, which mark the boundary between air masses of different temperatures. The lines with triangles represent cold fronts, while lines with hemispheres show warm fronts. Frontal systems usually bring weather changes (see p. 34).

... watching the weather,
accepting it, knowing all
things change ...

HAL BORLAND (1900–1978),
American naturalist

PUTTING IT ALL TOGETHER

Once you can read weather maps you will find forecasts more informative. Look for the major features to see what they are bringing to your region. You can then interpret what is being said about the movement of pressure systems and frontal systems and work out what weather conditions to expect.

The Amateur Weather-Watcher

Familiarizing yourself with local weather conditions is the first step in becoming a weather-watcher.

An interest in amateur meteorology can be supplemented by many of the techniques used by professional forecasters. There is also a wide range of meteorological instruments that can be used at home (see pp. 96–9). You may want to use your observations to confirm weather forecasts, or to make predictions of your own. However, the best way to begin to understand the weather is to simply gaze at the sky and follow its ever-changing display.

Signs in the Sky

Learning the names of clouds can be a delightful activity in itself (see pp. 188–217). By becoming familiar with cloud types and their movement, you will also be able to make educated guesses about developing weather conditions.

Most of us are aware that a dark, threatening sky means a rainstorm or snowstorm may be imminent, and that a steady decrease in the number of clouds usually means that conditions are improving.

The three main types of cloud to recognize are high-level, middle-level, and low-level (see p. 44). A build-up of high-level or middle-level clouds, such as cirrus or altocumulus, may indicate an approaching frontal system. When low-level clouds, such as stratus, move in behind middle- and high-level clouds, rain and snow may be coming.

Cumulus clouds form vertically, and may extend from low to high levels. Fair-weather cumulus clouds (cumulus humilis) usually indicate continuing settled weather. At the opposite extreme, cumulus clouds growing in size over the

OUTDOOR ACTIVITIES *are always enhanced by weather-watching. In some sports, such as hang gliding (above), an understanding of the weather is vital.*

IN THE CLOUDS *A steady build-up of altocumulus clouds (above) often heralds a frontal system. Cumulus congestus clouds (right) can produce heavy rainfall.*

NEAR AND FAR *Photography captures all sorts of weather phenomena, from ice crystals on a window (below) to lightning over Boulder City, Nevada, USA (bottom).*

PHOTOGRAPHING THE WEATHER

Weather phenomena are the ideal subjects for photography. By their very nature, they may be only fleeting events which have to be captured in an instant. It is a good idea to have your camera nearby so you can catch the moment.

There are two main categories of phenomena to record. The first group consists of large, distant, and dramatic features such as cloud formations, sunsets, thunderstorms and lightning, and rainbows, haloes, and other optical effects. At the other end of the spectrum, the minute—including hoar frost on plants or on windowpanes, snowflakes, and hailstones—can provide both exquisite images and reliable records of extreme weather events.

To record such widely different images effectively requires appropriate equipment. The most versatile camera to use is a 35 mm SLR model, which allows you to change lenses according to the subject. You can use a wide-angle lens to capture the full expanse of the sky. A macro lens or attachment allows you to magnify the subject and reveal its details. A telephoto lens can be used to photograph both close-up and distant views. For successful photographs of lightning, you may need to experiment with slow shutter speeds.

A good time to look for ice and dew details is early in the morning. The best shots of large features are taken out in the open country. Better still, mountains, tall buildings, and airplanes can all provide a view of the horizon, making it much easier to frame spectacular images.

course of a few hours indicate instability and the threat of thunderstorms that may bring fierce, gusty winds, strong updrafts, and heavy rain and hail.

In middle latitudes, the weather generally moves from west to east, but local winds may be influenced by other factors. For instance, in coastal locations, sea breezes may blow in a completely different direction from winds being experienced a few miles inland (see p. 32). Furthermore, as a large thunderstorm approaches, it draws air into it, so that the surface wind may blow toward the storm, in the opposite direction to the storm's movement. This explains the old weather saying that thunderstorms come up against the wind.

By observing the wind and monitoring a barometer (see p. 96), you can sometimes get an idea of the approaching weather. In the middle latitudes of the Northern Hemisphere, for example, falling air pressure accompanied by increasing wind from the southwest can indicate the approach of a frontal system.

HALOES *occasionally appear around the Sun (right) or the Moon. They are usually associated with cirrus clouds.*

Special Effects

Particular weather conditions produce optical phenomena that are both beautiful and informative. Rainbows are probably the best known of these phenomena, but there are many others to look out for. They are all created by the combination of sunlight or moonlight and particles in the atmosphere, and can sometimes tell you how the weather is likely to behave.

Haloes sometimes appear around the Sun or the Moon. They tend to be associated with cirrus clouds. These, in turn, often signal advancing cold fronts, which sometimes bring severe thunderstorms.

READING *the* SIGNS *in* NATURE

The state of the weather can be linked to the behavior of wildlife and the growth of plants and trees, so observing nature will enhance your weather knowledge.

Animals and plants respond to changes in the weather, and by watching them you will become familiar with their responses. Although they are certainly not reliable forecasting tools, you can gain a fascinating insight into the connections between nature and the weather.

WEATHER LORE

In folklore, the behavior of animals is often interpreted as foretelling a change in the weather, particularly the approach of rain. Nature does react to incoming storms in a number of ways: insects are more active, bees return to their hives, birds and bats fly lower, frogs croak more, flowers open and close—all in response to the changing air pressure and humidity.

Cows lying down or gathering together in a corner of a field have long been seen as a sign of rain to come. Often as a storm is developing, cows will lie down to keep a patch of grass dry or huddle together as a protective response.

Swallows and swifts flying high in search of insects are seen as a sign of temporarily settled weather with the risk of a thunderstorm. This is because on warm afternoons and evenings, rising air carries insects aloft and these rising currents are often associated with the sultry weather that may precede thunderstorms.

A number of sayings links the production of spiders' webs with calm, settled conditions and their absence with the probability of rain. There is little evidence to support these claims, although it is true that spiders will not spin webs during heavy rain because of the damage it can do to their handiwork.

The link between weather and the emergence of leaves and flowers from dormant buds is the subject of much analysis and speculation (see box). Many plants respond to changes in humidity, which

WEATHER PREDICTIONS *based on folklore often contain an element of truth. Pine cones (left) open and close in response to changes in humidity. Cows don't like to lie on wet grass so they lie down before rain (below), while spiders, such as this black-and-yellow argiope (right), will stop weaving their webs during heavy rain.*

FOLKLORE *in North America has it that if a groundhog (left) sees its shadow at noon on 2 February, the following six weeks will be cold.*

FIRST FLOWERING

The study of the progress of vegetation through the seasons, known as phenology, is a fascinating subject. The Swedish naturalist, Carolus Linnaeus (1707–78), was the first to outline a method of recording leaf-opening, flowering, fruiting, and leaf-fall in order to create climatological records.

One of the longest phenological records was kept by the Marsham family in Norfolk, England. Starting with Robert Marsham, who kept a diary from 1736, six generations of the family kept records until 1947. These included the first flowering of 4 species (snowdrop, wood anemone, hawthorn, and turnip), the first leafing of 13 kinds of tree, and 10 other dates concerning birds, butterflies, and frogs.

Phenological records provide important data about the weather before the advent of modern instrumental observations (see p. 114). English records show that the date when some native trees came into leaf has varied by up to three months between the mildest and coldest springs. This is a measure of the adaptability of plants and an example of the intriguing results such records can yield.

THE BLOOMING *of the snowdrop (right) each spring was recorded by six generations of the Marsham family.*

Oak before Ash

We'll only have a splash

Ash before Oak

We're in for a soak.

English proverb

may, in turn, indicate the likelihood of rain. If the air is dry, the chance of rain is considered low.

In England, the scarlet pimpernel is seen as the poor person's weather glass because its flowers close when humidity is high. Pine cones also close during humid weather. Seaweed becomes limp when the air is humid, and rigid when it is dry.

Seasonal Signs

The return of migrating birds each year is the first sign of spring in many parts of the world. In North America, the returns of the bluebird and the robin are eagerly awaited for this reason. The eighteenth-century English naturalist Gilbert White kept records of the arrival of the swallow for 40 years. The timing of the swallow's arrival in Britain is still widely regarded as a sign of what the summer is going to be like—an early arrival is said to mean a dry summer. The same is said of the cuckoo, while the early arrival of the waxwing in fall has been seen as a sign of a harsh winter. In fact, these movements are more indicative of the weather the birds have left behind than of future weather.

Landscape

Natural landforms can help you identify weather conditions. Take note of the trees, hills, and buildings around where you live. How far you can see at any given time is a rough measure of how clear the air is. In general, the higher the humidity, the lower the visibility, and vice versa.

VISIBILITY *is a good indication of the level of humidity in the air. Mist or fog (below) will greatly reduce visibility.*

A SIGN OF SPRING *in North America is the arrival of the bluebird.*

Observing *at* Home

With a few basic instruments and a methodical approach, you can compile a fascinating record of the local weather and its effects on your life.

BEAUFORT SCALE				
Code	**Speed mph**	**Speed kph**	**Description**	**Effects on land**
0	below 1	below 1	calm	smoke rises vertically
1	2–3	1–5	light air	smoke drifts slowly
2	4–7	6–11	light breeze	leaves rustle; vanes begin to move
3	8–12	12–19	gentle breeze	leaves and twigs move
4	13–18	20–29	moderate breeze	small branches move; dust blown about
5	19–24	30–38	fresh breeze	small trees sway
6	25–31	39–51	strong breeze	large branches sway; utility wires whistle
7	32–38	51–61	near gale	trees sway; difficult to walk against wind
8	39–46	62–74	gale	twigs snap off trees
9	47–54	75–86	strong gale	branches break; minor structural damage
10	55–63	87–101	whole gale	trees uprooted; significant structural damage
11	64–74	102–120	storm	widespread damage
12	above 74	above 120	hurricane	widespread destruction

The first question to address when deciding to set up your own weather station is: how serious do you want to be? There is a huge range of instruments available to measure the weather, but you can start with a just a few simple ones (see p. 96). Some observations require no instruments at all. For instance, you can estimate how much of the sky is covered by clouds, and wind speed can be estimated using the Beaufort scale (see table).

What you can achieve will depend partly on where you live. If you cannot site instruments well clear of nearby buildings, you will have to accept that your readings, while providing a useful set of localized observations, may not be an accurate measure of the climate.

Keeping Records

Once you have set up your instruments, you may want to keep records of day-to-day fluctuations. Whatever measurements you take at home can be supplemented by media observations, but you will have to be discriminating if you do not want to be overwhelmed by data. In deciding what to record, remember that regular, methodical readings will be much more useful over time than occasional records of extreme events.

INSTRUMENT READINGS *(left) should be made twice daily, preferably in the early morning and afternoon. These readings, and other weather observations, can then be kept in a diary (top left).*

Most national weather services have standard weather sheets for recording daily conditions. These usually have columns for at least the following: sky condition (clear or the amount of cloud cover, its height and type); wind speed and direction; visibility; dry-bulb and wet-bulb, and maximum and minimum temperatures; minimum grass and soil temperatures at various depths; snow depth and new snowfall; amount and type of rainfall in the last 24 hours; total sunshine for the day; and your comments

EXTREME WEATHER, *such as a fall of hail (below), may be exciting, but the most useful records are those kept regularly over a long period of time.*

about weather developments over the preceding 24 hours.

You may decide to make a simpler record sheet, covering only those measurements you make regularly. A diary, on the other hand, provides plenty of space for personal observations. You might want to describe unusual weather events such as lightning, hail, snowfall, and fog, as well as record the impact of frost or drought on your garden, which animals become more active before a storm, or the arrival of migrating birds.

Becoming familiar with the standard weather symbols (see p. 85) will provide you with a quick, graphic way to record the weather. Sketching cloud formations is an excellent way to improve your powers of observation, as is taking photographs (see p. 91).

FAMILY RECORDS

A wealth of meteorological information has been collected by volunteers. For instance, one of the oldest weather records in the eastern states of Australia has been compiled at a property called Buckalong in southeastern New South Wales. For the past 137 years, the owners of this property have kept uninterrupted daily rainfall records. The present owner, Tony Garnock (right), has made these daily readings for the last 50 of these years. Every morning at 9 am he checks his home weather station, recording any rainfall on a form that he then sends to the Bureau of Meteorology in Melbourne.

Continuous rainfall records, such as those kept by the Garnock family, provide information about local long-term rainfall patterns that are invaluable to farmers trying to plan for the following season.

Making Forecasts

In spite of the high quality of current forecasts, your own observations can provide additional insights into how the local weather may develop. When the overnight temperature is forecast to fall close to freezing, local conditions will then determine whether a killing frost will develop and ice will form on the roads. While forecasters have some difficulty predicting the precise path of winter snowstorms, if you monitor the temperatures closely you may be able to decide whether the snow will continue or turn to rain. When major storms are approaching, try to keep an eye on the barometer. The rate at which the air pressure is falling will provide clues as to how the weather is matching up to the forecast.

The Next Step

If you become increasingly interested in the weather and want to exchange information or meet with other weather enthusiasts, the first step is to find out about the local chapters of national meteorological societies (see p. 274). Another way to find fellow enthusiasts is through computer networks such as the Internet.

If you become completely hooked on recording the weather, and have been making reliable measurements for at least two or three years, you may be able to become an official observer (see p. 82). However, you will need to have an appropriate set of instruments and live at a site that fills a gap in the local network. Your national weather service will tell you whether you can provide a service they need.

PERSONAL COMPUTERS *add another dimension to home weather observations. Software programs allow you to make graphs from your own readings, and you can gain access to a wealth of weather information on the Internet.*

MEASURING *the* WEATHER

Both professionals and amateurs use a wide range of instruments and equipment to monitor the ever-changing behavior of the weather.

BAROMETERS *An aneroid model from the eighteenth century (left); a mercury barometer in use today (below left).*

To understand the weather and predict its behavior, we need to know the properties of the atmosphere. These include the air temperature, air pressure, wind speed, relative humidity, and cloudiness. In addition, it helps to know what the visibility is and how much rain or snow has fallen. These parameters can be measured using a variety of instruments, from home-made rain gauges to orbiting satellites.

When setting up a home weather station, it is best to begin with just a few instruments, such as an aneroid barometer, simple maximum and minimum thermometers, and a basic rain gauge. From these modest beginnings you can progress, in time, to keeping an extensive computer-controlled set of instruments that will record the complete range of meteorological parameters.

AIR PRESSURE

Fluctuations in the air pressure often herald weather changes (see p. 26). Rapidly falling pressure, for example, indicates that a low-pressure system is approaching, which may bring rain. A standard aneroid barometer—the most widely used instrument for measuring air pressure—should be part of any weather-watcher's kit.

An aneroid barometer consists of a vacuum capsule, made of steel or beryllium. When the air pressure rises, the capsule is compressed; when the pressure falls, the capsule expands. The capsule movement is translated by a set of levers into the movement of a pointer. In most barometers, this pointer is viewed on a dial. In barographs, it is connected to a pen that plots the pressure changes on cylindrical graph paper.

It is generally best to keep your barometer indoors, away from sunlight and drafts. If the cylinder is markedly heated or chilled, readings are distorted.

You can use a barometer to check the progress of high- and low-pressure zones across your local area. This can also be done automatically using either a barograph or computerized systems. As air pressure varies with altitude (see p. 26), barometric readings are usually converted to sea level. You may want to obtain a conversion table from your weather office so that you can compare your readings to those in the media.

Aneroid barometers are liable to drift a little with time and need regular recalibration to ensure accurate readings. Absolute measurements of air

A TERRESTRIAL THERMOMETER *can be used to measure air temperatures at ground level, which usually differ from those taken in an instrument shelter.*

TEMPERATURE CONVERSIONS

From Celsius to Fahrenheit: °F = (1.8 × °C) + 32
From Fahrenheit to Celsius: °C = 0.56 × (°F – 32)

ELECTRONIC DEVICES, *such as the Weather Wizard (below), can be used to monitor several parameters—from temperature to wind speed.*

pressure are made using a barometer that measures the height of a mercury column in a vacuum (see p. 67). This is the most accurate way of measuring pressure, but it has been largely phased out because of the convenience of aneroid barometers.

Air pressure is measured in hectopascals (until recently known as millibars) or, less commonly, in inches or millimeters of mercury. Modern precision instruments can give readings to an accuracy of about 0.1 hectopascal.

TEMPERATURE

Most thermometers contain either mercury or alcohol—substances that expand when they are heated and contract when cooled. Maximum and minimum thermometers have metal indicators that record the highest and lowest temperatures in an observation period.

Thermographs are recording thermometers usually made from two metal strips welded together. As the temperature changes, the metals expand by differing amounts, causing distortion. The movement of the strips is translated into temperature movement and recorded as on a barograph.

To obtain a true reading of the air temperature, you should place your thermometer in an instrument shelter (see box) or a permanently shaded area. Minimum temperatures are the best to use for day-to-day comparisons because they normally occur when there is no sunlight, either direct or indirect, to affect the situation.

A good check of your readings is to compare them with those made at the nearest official recording site—often the local airport. Providing the area where you live is not much more built up, or at a significantly different altitude, the figures should not differ by more than a degree or two.

In the United States, temperature is generally measured in degrees Fahrenheit. In most other countries, the Celsius scale has been adopted (see p. 68).

THE INSTRUMENT SHELTER

The measurement of air temperature has always presented a challenge. Plainly, it is much hotter out in the midday sun, and we may feel that when meteorologists talk about the shade temperature they are underestimating the actual temperature. However, temperatures in the sun are influenced by nearby surfaces and are of limited use because different surfaces absorb and radiate different amounts of heat—a tarmac road will become much hotter than white walls.

To obtain a meaningful reading of the air temperature, thermometers must not be affected by direct sunlight or radiation from surfaces heated by sunlight. Meteorologists therefore place their instruments in a shelter known as a Stevenson screen or weather shack. This is a louvered box that can hold a collection of instruments at a height of 4 feet (1.2 m) above the ground. A fully equipped shelter may contain a barometer, maximum and minimum thermometers, a thermograph, wet-bulb and dry-bulb thermometers, and a hygrograph (see p. 99).

A standard shelter has slatted sides to prevent sunlight and heat radiation reaching the thermometers, while permitting a free flow of air past the instruments. The roof is double-layered to prevent sunlight heating up the interior. The whole shelter is painted with white gloss paint and preferably situated above ground covered by short grass and at least 32 feet (10 m) from any buildings.

A THERMOGRAPH *(below) is a form of thermometer that plots temperature readings against time, in the same way as a barograph records air pressure.*

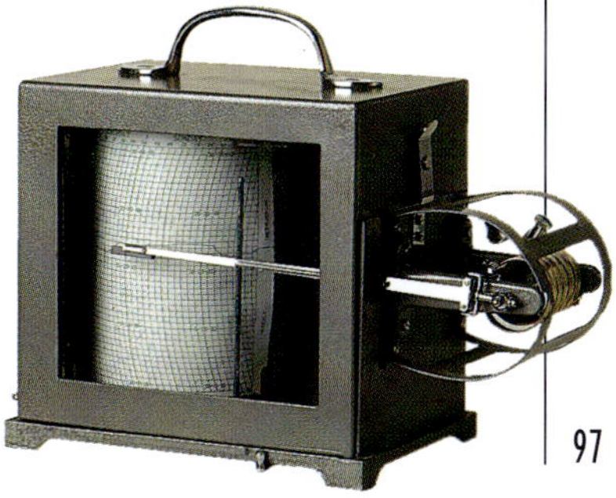

Rain and Snow

Any flat-bottomed transparent container with straight sides can be used to make a rain gauge. A more effective gauge, however, can be made by putting the stem of a funnel in a narrow-necked jar or bottle, as this will prevent the evaporation that otherwise occurs. Rainfall is measured in inches or millimeters.

Rainfall measurements using a rain gauge are affected by wind patterns and eddies created by buildings, trees, and the gauge itself. But given how greatly rainfall can vary from place to place, there is no substitute for taking your own measurements to keep track of local conditions. Even if your instrument produces slightly inaccurate figures, they will be good enough for most purposes. For instance, gardeners need to be aware of variations in rainfall. Knowing how much rain has fallen in a dry spell helps them decide how much watering sensitive plants will require.

RAINFALL *The best rain gauges have a funnel-shaped opening (left). Pluviographs (below) plot rainfall recordings on a rotating drum over a period of time.*

A rain gauge is of limited use when snow falls. The easiest way to measure snow cover is to push a ruler into the snow until it reaches the ground.

Following the Wind

To measure wind speeds, you will need a cup anemometer. This device consists of four metal or plastic cups on arms that are mounted perpendicular to a central rotating shaft. The stronger the wind, the more rapidly the cups will rotate on the shaft. In the simplest models, the number of rotations is counted by the observer. Some models display the number of rotations on a scale. There are also electronic anemometers, which provide a continuous record of wind speed.

An anemometer needs to be installed in a reasonably open space or high enough to avoid the distorting effects of nearby buildings and other obstructions. If there is no convenient place to mount it, you can use a hand-held instrument. Alternatively, you can rely on general indicators of wind speed established as part of the Beaufort scale (see p. 94).

Wind vanes and wind socks indicate wind direction, and wind socks also give some idea of wind speed. These should also be situated away from trees and buildings.

A SUNSHINE RECORDER *being checked in the Tanami Desert, Australia.*

WIND DIRECTION *is indicated by wind vanes (left) or wind socks (below). The sock also helps estimate wind speed: the straighter the sock, the stronger the wind.*

Sunshine and Cloud

The hours of sunlight in a day can be measured using a sunshine recorder, which focuses the Sun's rays through a glass ball onto a strip of card. As the Sun travels across the sky over the course of a day, it leaves a trail of scorch marks on the card, but if cloud intervenes, the trail is broken.

From the ground, there is no accurate way to measure the extent of cloud cover, so forecasters rely largely on satellite images for this information (see p. 100). For your own records, however, it is worth making a rough estimate of the extent and types of clouds that you can see. You could use the standard symbols to do this (see p. 85).

Humidity

To accurately measure humidity (the amount of moisture in the air) you will need wet-bulb and dry-bulb thermometers. Humidity varies with temperature, and warm air can hold considerably more moisture than cold air. This variation means that we do not normally measure the absolute moisture levels but rather the amount relative to the maximum that the air can hold at a given temperature—the relative humidity—which is expressed as a percentage (see p. 40).

The wet-bulb thermometer is a well-ventilated thermometer with the bulb surrounded by wet muslin. A comparison is made of the temperatures recorded by the wet-bulb and dry-bulb thermometers, thus comparing the temperature of the saturated air with the actual air temperature. If the wet-bulb and dry-bulb readings are the same, then the air must be saturated with moisture and have a relative humidity of 100 percent. This means that the air has reached its dewpoint, and dew (see p. 180) or frost (see p. 186) is likely to occur overnight—useful information if you are a keen gardener. You can convert your readings to relative humidity percentages by using standard tables, which are usually supplied along with the thermometers.

Hygrometers and hygrographs are other instruments used to measure humidity. On hygrometers, the relative humidity is displayed on a dial; on hygrographs, it is recorded graphically. Hygrometers and hygrographs work well at high temperatures and humidities but become increasingly inaccurate as temperatures and humidity levels drop.

Human hair expands in moist air and contracts when it is dry, so some hygrometers and hygrographs contain a set of hairs under tension. Variations in the length of the hairs are registered as relative humidity changes.

WIND SPEED *is best measured by using a cup anemometer. If there is no convenient place to mount one, a hand-held instrument can be used (above). Hygrographs (right) are used to record humidity changes. This model uses strands of human hair.*

Electronic Instruments

Technological developments this century have enabled weather forecasters to gather a vast range of information about the Earth and its atmosphere.

Many of the instruments that we use to measure the weather have been based on the same design for centuries. In recent decades, however, meteorology has advanced at an astonishing pace, thanks to the development of a range of electronic instruments.

Electronic sensors are able to measure weather conditions as accurately as standard meteorological instruments, and are used extensively in remote locations, in ocean buoys, and on aircraft.

At some sites, observers launch balloons carrying radiosondes, which are small packages of electronic instruments. These devices measure temperature, air pressure, and humidity as they rise through the atmosphere to a height of about 100,000 feet (30,000 m), and transmit the information back to surface stations. Radar systems are used to track the paths of radiosondes and thus gather information about wind speed and direction in the upper atmosphere.

Radar and Rainfall

Ground-based radar equipment is used to determine how much rain or snow is falling and where it is heaviest. Rain and snow scatter radio signals, so by sending out pulses of radiation from a transmitter and then measuring how much of the signal is reflected back to a receiver,

REMOTE RECORDING *The Meteosat weather satellite (top) produces infrared images of the Earth's atmosphere every 30 minutes, and also acts as a data relay. Some automatic weather stations (below) transmit their readings to satellites, which send them on to weather agencies.*

RADIOSONDE BALLOONS *(right) are sometimes filled with hydrogen, an explosive gas, so people handling them may need to wear protective clothing.*

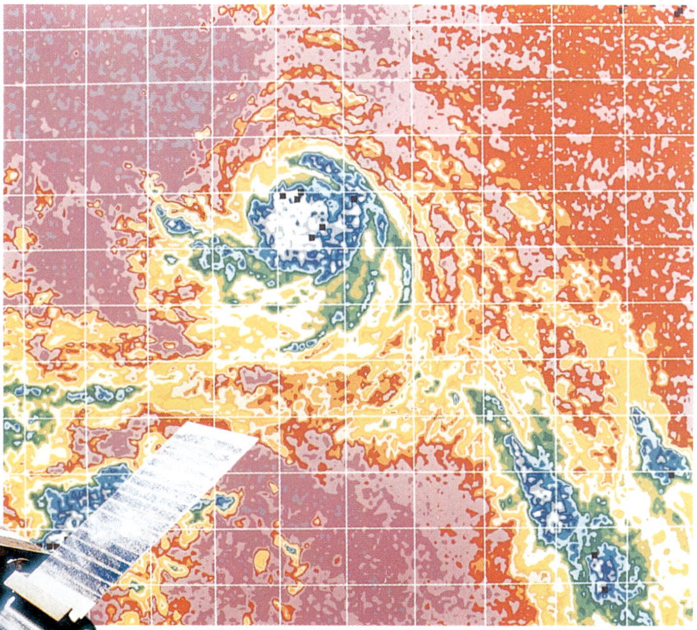

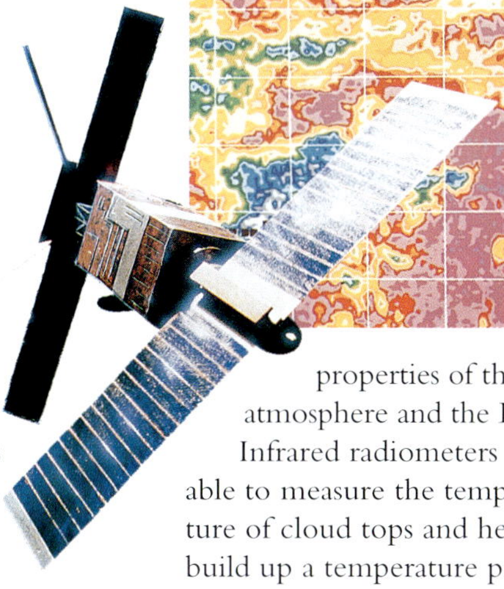

THE VIEW FROM SPACE *An early example of the worth of satellites was the monitoring of Hurricane Camille in 1969 (right). The European Remote-Sensing Satellite, ERS-1 (below), was launched in 1991, primarily to study coastal, ocean, and polar-ice processes.*

it is possible to create a detailed image of the pattern of precipitation.

Radar images are most effective in plotting the development and movement of the heavy rainfall associated with severe thunderstorms. The reflected signals can be converted into rainfall rates, which assist those involved in flood control and the management of water resources. Doppler radars (see p. 104) have been used to improve forecasts of damaging winds, such as tornadoes.

WEATHER SATELLITES

Several weather satellites orbit the Earth, producing both standard photographs and infrared images. Some also carry radar equipment and radiometers, which can be used to measure various properties of the atmosphere and the Earth.

Infrared radiometers are able to measure the temperature of cloud tops and help build up a temperature profile of the atmosphere. Microwave radiometers have the huge advantage of being able to "see" through clouds. They not only provide thorough atmospheric temperature measurements, but also observe the extent of snow cover and sea ice in polar regions, which are frequently shrouded in clouds.

For over 15 years, a range of instruments on the United States' National Oceanic and Atmospheric Administration (NOAA) weather satellites has been employed to monitor temperatures and snow and ice cover. These records are a vital component in tracking the progress of global warming (see p. 126).

The European satellites ERS-1 and ERS-2 carry radar and microwave equipment that can measure ocean wave heights, wind speeds, changes in the levels of the polar ice caps, and the structure and strength of ocean currents.

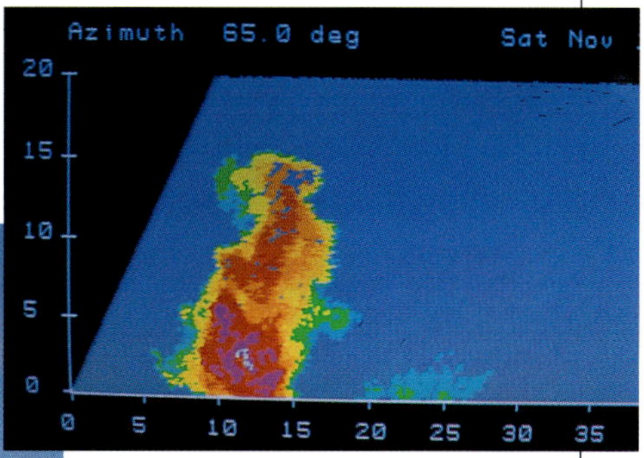

STORM TRACKING *Meteorological radar stations (left) help provide precise, short-term forecasts of severe thunderstorms and tornadoes. This radar image (above) shows a thunderstorm. Radar is also used to track radiosonde balloons, which provides data about the speed and direction of upper-atmosphere winds.*

Seasonal Forecasts

There is growing evidence that long-range forecasts can provide a reasonable guide as to whether temperature and rainfall will be above or below average in the weeks and months ahead.

LASTING CLIMATIC EFFECTS *are produced by slowly varying components, including sea-surface temperatures (left, in mapped form) and the extent of ice cover in the Arctic and Antarctic (right).*

Accurate seasonal forecasts are of great value to farmers when they are deciding which crops to plant, to energy utilities when they are planning load levels, and to the average person in organizing outdoor activities, such as a hiking vacation.

The Climate's Memory

The fundamental limit to forecasting the day-to-day weather is 10 to 14 days (see p. 105), but longer-term patterns may be predictable. The assumption implicit in much folklore is that past patterns of weather, as reflected in the responses of animals and plants, contain clues about future patterns. Many statistics support the hypothesis that the climate has a "memory", with slowly varying components—notably sea-surface temperatures (SSTs) throughout the world, and the extent of pack ice in polar regions—producing effects that last months or even years.

El Niño

There is mounting evidence that changes in the surface temperature of the tropical Pacific Ocean are responsible for controlling many features of global weather patterns. Such changes have been linked to El Niño, a warm ocean current that appears off the northwestern coast of Peru around Christmas. In some years, it flows much farther south than usual, creating what is known as an El Niño event. The warm water, lacking in nutrients, spreads over the top of cold, nutrient-rich water and kills marine life, which leads to the death and migration of birds and economic disaster for local people. The current may also trigger heavy rains that cause flooding and erosion.

El Niño events occur every two to seven years and can

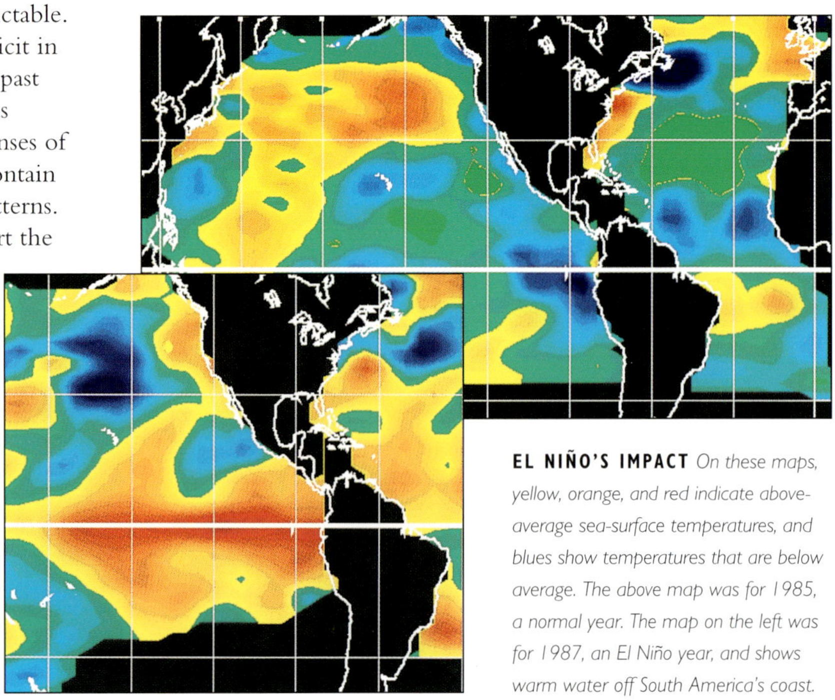

EL NIÑO'S IMPACT *On these maps, yellow, orange, and red indicate above-average sea-surface temperatures, and blues show temperatures that are below average. The above map was for 1985, a normal year. The map on the left was for 1987, an El Niño year, and shows warm water off South America's coast.*

last up to three or four years. We now know that they affect the Earth's trade-wind patterns (see p. 30), which, in turn, influence SSTs over vast areas of the Pacific. These changes can produce extreme weather throughout the tropics and have been linked to severe droughts in Australia. They also affect the weather farther afield, causing anomalies in winter temperatures over North America and in winter rainfall over northwestern Europe.

FLOODS AND DROUGHT *During El Niño years, flooding often occurs in South America (above). A disastrous drought in Australia in 1982 and 1983 (right) has been attributed to the El Niño event of the same years.*

Predicting Patterns

As we can predict El Niño events several months ahead, the potential of basing long-range forecasts on El Niño has been examined. Using a computer model, the British Meteorological Office simulated rainfall in northeastern Brazil on the basis of historical records of tropical SSTs. In 93 percent of cases, their results matched the actual rainfall records.

From Peru to Zimbabwe, forecasts based on SSTs are already being used to make decisions about which crops to grow each rainy season. Outside tropical regions, however, forecasts using the El Niño patterns are proving less reliable. Nevertheless, the United States National Weather Service has decided to extend its seasonal forecasts out to beyond a year.

To date, it seems that, in middle latitudes, forecasts for winter and summer have some value, but those for spring and fall are virtually worthless. This result is not unexpected, as during winter and summer the patterns settle into the two extremes of the annual cycle and become more susceptible to the influences of slowly varying components of the climate system, such as SSTs. By contrast, during the transitional periods of spring and fall, the chaos of short-term changes is likely to swamp these underlying influences.

Although seasonal forecasts may offer only slightly better odds on one type of season occurring, this information might still prove valuable. For example, farmers could decide to plant crops on the basis of the next season being dry or wet. Similarly, energy utilities could plan with some confidence on winter demand being above or below average. Seasonal predictions may well become as much a part of our lives as the standard daily forecasts.

LONG-RANGE FORECASTS *based on El Niño predictions are proving useful in the tropics (left). Improved seasonal forecasts will also be of interest to vacationers (below).*

THE FUTURE *of* FORECASTING

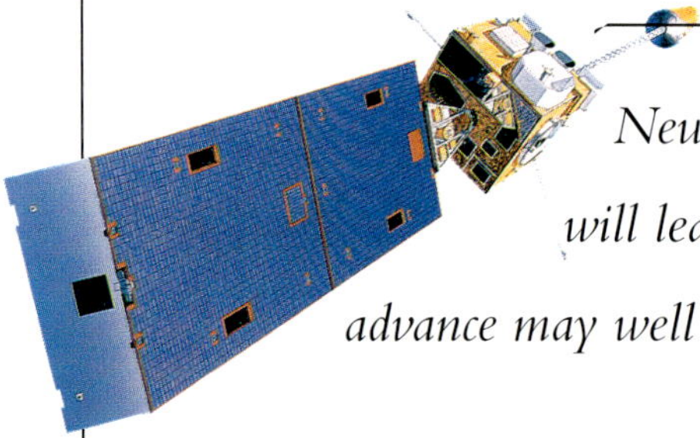

New technology and increasingly powerful computers will lead to improved predictions, but the major advance may well be in our understanding of forecasting's limits.

Contrary to popular belief, weather forecasting has improved greatly in the last couple of decades, mainly because of a vast increase in computer power. This, combined with satellite measurements and a better understanding of how the weather works, has enabled forecasters to double the range of useful predictions since the early 1970s. On average, forecasts are now good for just over six days, with winter predictions about a day better and summer ones falling about a day short of the average.

THE FIRST IN THE SERIES *of GOES satellites (above) was launched in 1994. This joint project of NOAA and NASA in the USA plans to add four more satellites by 2003. GOES satellites are geostationary, which means that they remain over the same point on the Earth's surface.*

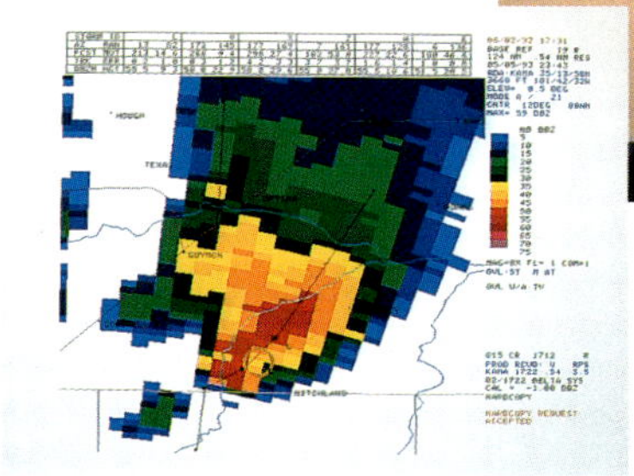

TORNADOES *(above) can be effectively tracked by Doppler radar. In this Doppler image (left) the tornado is shown in red.*

RADAR AND SATELLITE

Increasingly sensitive radar systems are helping to improve forecasts for a few hours ahead. In the United States, where 10,000 violent thunderstorms, 5,000 floods, and nearly 1,000 tornadoes strike each year, the National Weather Service has installed a Doppler radar system that can peer deep into thunderstorms and detect when they are becoming dangerous. The successful identification of damaging storms has risen sharply, the number of false alarms has dropped to less than half of the previous figure, and warning times have increased.

Better measurements, particularly from weather satellites, and the use of more sophisticated computers will also improve forecasts for a few days ahead. Countries developing more advanced satellite systems include the European Community countries (as part of the European Satellite Agency), Japan, the United States, India, and Russia.

DOPPLER RADAR *(left) uses two radars to provide a "stereographic view" of changes in signal frequency and can track small-scale changes within a storm.*

ENSEMBLE FORECASTS

Despite the vast amounts of data collected daily, the atmosphere is so complex that our knowledge of its state at any given moment will probably always be imperfect. How much these slight inaccuracies affect the reliability of forecasts depends on whether the weather worldwide is in a quasistationary state, which makes it relatively predictable, or whether the regime is about to break down in a rapid and unpredictable manner.

READY TO ORBIT *ERS-2, the latest European satellite, was launched in 1995 to complement the observations of its twin, ERS-1 (see p. 101).*

A method known as ensemble forecasting has been developed to help determine whether the weather is in a predictable phase. It involves running a computer simulation several times, each time entering slightly different initial conditions for the atmosphere. If the ensemble of forecasts looks remarkably similar up to 10 days ahead, then there is a good chance the forecasts will be reasonably accurate. If, however, each forecast diverges after a few days, then the atmosphere is in a less predictable mood.

The principal forecasting bodies (see p. 80) are all experimenting with ensemble methods. In the future, forecasts for about a week ahead will place greater stress on the probability of the prediction happening. Sometimes, forecasters will have to admit that they are unable to predict what is going to happen beyond a few days, while at other times they will be confident enough to provide forecasts for up to 10 days ahead.

FURTHER AHEAD

We may never be able to produce detailed forecasts further than 10 days ahead, but general outlooks up to a few months ahead should become more accurate. Such improvements will depend on finding out more about why the weather in middle latitudes becomes stuck in certain patterns (see p. 117). Each summer and winter, one of a limited number of well-defined patterns tends to predominate for weeks on end. The balance of these patterns throughout the season determines whether temperatures and rainfall are above or below normal.

CHAOS THEORY

Chaos theory has its roots in meteorology. Edward Lorenz's work at the Massachusetts Institute of Technology in the early 1960s provided insights into systems with nonlinear properties. In such systems, as one parameter changes, others alter in a way that is not in direct proportion to this change. Using a set of equations to represent convection processes in the atmosphere, Lorenz showed that its behavior was inherently unpredictable. The system never returned to precisely the same state and so could never repeat itself.

This conclusion has profound implications for weather forecasting. Although the weather may follow broadly similar patterns over the years, and these define the climate, it will never precisely repeat past patterns. Furthermore, a tiny disturbance in the air can trigger a major weather occurrence elsewhere at a later time because of the interconnectedness of the atmosphere. This is described as the "butterfly effect", because, in theory, a butterfly stirring the air in Beijing today can transform storm systems next month in New York.

We can never precisely define the current state of the atmosphere, and the butterfly effect suggests that even tiny errors in our models will grow until they completely overwhelm the calculations. In practical terms, this means that, even with greatly improved measurements and increased computing power, the growing errors in our calculations will usually overwhelm the daily forecasts beyond 10 days ahead.

EDWARD LORENZ *(b. 1917) did pioneering work on chaos theory.*

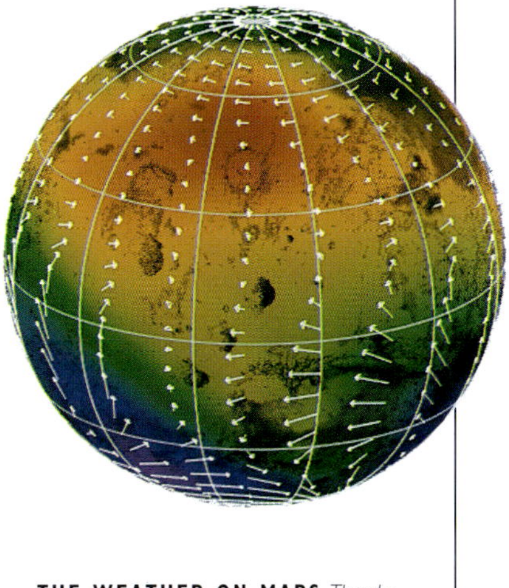

THE WEATHER ON MARS *Thanks to NASA, those of us who are interested in off-world activities can now use the World Wide Web computer network to check the weather on Mars (above).*

The weather is always doing something … always getting up new designs and trying them on the people to see how they will go.

MARK TWAIN (1835–1910),
American writer and humorist

CHAPTER FIVE
CHANGING WEATHER

A History *of* Climate Change

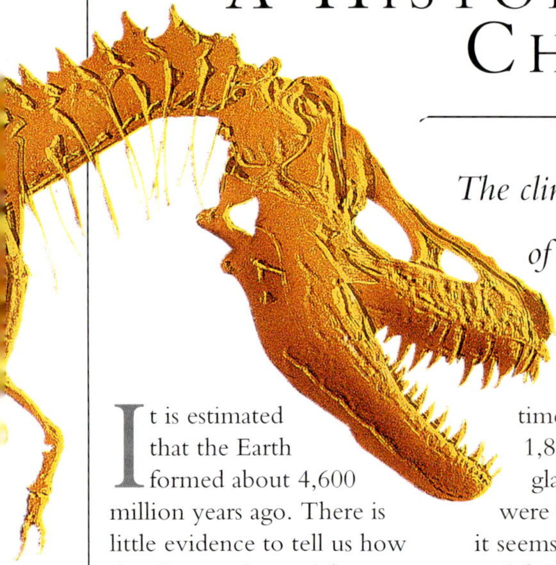

The climate has changed throughout every period of the Earth's history, and fossil records provide the evidence of these changes.

It is estimated that the Earth formed about 4,600 million years ago. There is little evidence to tell us how the climate changed for some 90 percent of the Earth's lifetime. We do not know where the oceans and the continents were, or what precisely the constituents of the atmosphere were.

The first sedimentary rocks were laid down some 3,700 million years ago in the Precambrian Era, when, it is believed, the climate was some 18° F (10° C) warmer than now. Algae, the first forms of life, appeared about 3,500 million years ago, but their fossils provide little evidence of climate change. All we know is that at some time between 2,700 and 1,800 million years ago, glaciers and ice sheets were widespread. Thereafter, it seems the Earth was warm and free of ice caps and glaciers for 800 million years.

The Late Precambrian Period began about 1,000 million years ago and included three distinct glacial episodes, each lasting some 100 million years. It is still unclear where these events occurred and how widespread they were, but there is no doubt they were major climatic events.

The Paleozoic and Mesozoic Eras

Following the Precambrian Era, the climate warmed appreciably and remained relatively warm for most of the following 300 million years. This era, from 570 to 245 million years ago, is known as the Paleozoic. Evidence of a brief glacial period about 450 million years ago can be found in the rock formations of the Sahara.

During the following period, the Carboniferous, temperatures dropped, culminating in the long Permo-Carboniferous glaciation from 330 to 245 million years ago. This icy epoch coincided with the formation of the super-continent Pangea, when all the Earth's landmasses came together and extended from pole to pole. What are now Antarctica,

METEORS *created huge craters (enormous versions of the one shown here). Some may have thrown enough dust into the air to block out the Sun, causing temperatures to fall—a cooling that could have wiped out the dinosaurs.*

FOSSILS *from earlier climates include:* Tarbosaurus bataar, *a Cretaceous dinosaur (left); crinoids, sea creatures from the Carboniferous Period (above); and fish from the Eocene Epoch (right).*

Continental drift over time

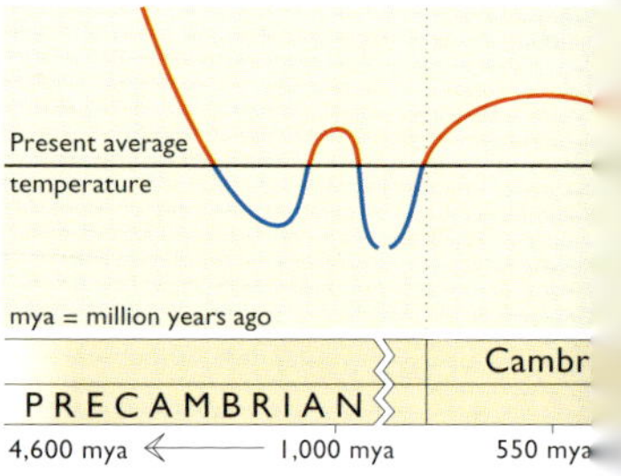

Australia, and India were located at high latitudes and formed the center of glaciation.

During the Mesozoic Era, from 245 to 65 million years ago, Pangea broke up into two enormous landmasses. The climate was generally warm, with relatively little difference in temperatures between the poles and the tropics—this was the age of the dinosaurs. There is evidence of some fluctuations in climate throughout this era, and there may have been a sudden, brief cooling at the end of the era that coincided with the mass extinction of the dinosaurs.

THE CENOZOIC ERA

The Cenozoic covers the last 65 million years of the Earth's history. The era involved a long-term cooling trend, although it was not a smooth decline: long, relatively stable periods were interspersed with more rapid, major cooling events about 50 and 38 million years ago. Further cooling about 15 million years ago led to the formation of mountain glaciers in the Northern Hemisphere, as well as the Antarctic ice sheet.

The second period of the Cenozoic Era is the Quaternary, which started about 1.6 million years ago and includes the present time. This period began with the Pleistocene Epoch during which there were seven glaciations, with up to 32 percent of the Earth's surface covered by ice. These ice ages occurred roughly every 100,000 years and were interspersed with shorter, warm interglacials.

The most recent glacial period reached its peak about 18,000 years ago. Ice sheets up to 10,000 feet (3 km) thick covered most of North America and all of Scandinavia, and extended to the northern half of the British Isles and the Urals. In the Southern Hemisphere, much of New Zealand and Argentina were under ice, as were the Snowy Mountains of Australia and the Drakensbergs in South Africa. Then, about 12,000 years ago, a dramatic warming began and by 7,000 years ago, the North American and Scandinavian ice sheets had melted. As sea levels rose, the coastlines of the continents gradually assumed their present shape.

We are living in the Holocene Epoch, a warm period that began 10,000 years ago. But we cannot be sure that the periodic glaciations have ceased, and we may well be in an interglacial period that will end with another ice age.

GLACIERS, *such as the LeConte Glacier in Alaska (above), are now mostly limited to high latitudes or altitudes. The Earth has, however, experienced many periods when glaciers were much more extensive.*

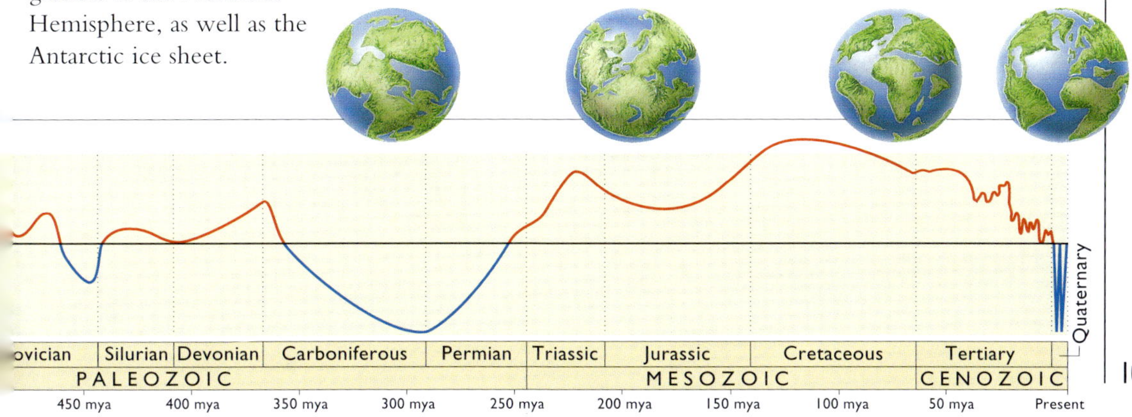

The Human Era

Following the cataclysmic fluctuations of the prehistoric era, a relatively stable period allowed human civilization to emerge and flourish.

A WARM CLIMATE *6,000 years ago fostered the birth of agriculture in Egypt (top left). At this time, the Sahara was far more fertile than today and supported large herds of animals, as evidenced by the Tassili N'Ajjer frescoes of Algeria (left).*

Human civilization emerged in a time of warm, stable weather, following a period of frequent glaciation and climatic turmoil (see p. 109).

About 6,000 years ago, the average temperature was some 4° F (2° C) higher than today and there was greater rainfall. These conditions fostered the birth of agriculture in Egypt and Mesopotamia, which led to surplus food stocks being kept for the first time. This enabled large groups of people to live together, and the first cities arose on the Mesopotamian plain.

The role climate played in the classical era, during the time of Greece's cultural flowering and the rise and fall of the Roman Empire, is still in dispute. The climate started to cool and many records of the time attribute crop failures to unseasonal weather conditions. The crop failures may, however, simply have been the consequence of deforestation and inefficient irrigation systems.

There are also complex reasons for the sudden decline of the Mayan civilization in Central America between AD 800 and 1000. The society was already under stress from population growth, environmental degradation, and intercity conflict, but it had overcome similar problems in earlier centuries. About AD 800, however, there was a long period of drought that would certainly have placed additional stress on the Maya, and may have caused the final collapse.

The Middle Ages

Broadly speaking, the tenth to twelfth centuries featured what is called the medieval climatic optimum, which, in Europe, consisted of temperatures similar to those today.

This favorable climate assisted the colonization of Iceland and Greenland by the Vikings, as well as the burgeoning of European civilization between the eleventh and thirteenth centuries.

As the thirteenth century came to a close, the climate changed again. There were dreadful cold, wet summers in 1315 and 1316 and a succession of cold summers thereafter. The failure of the Norse colony in Greenland and

Extent of glaciation

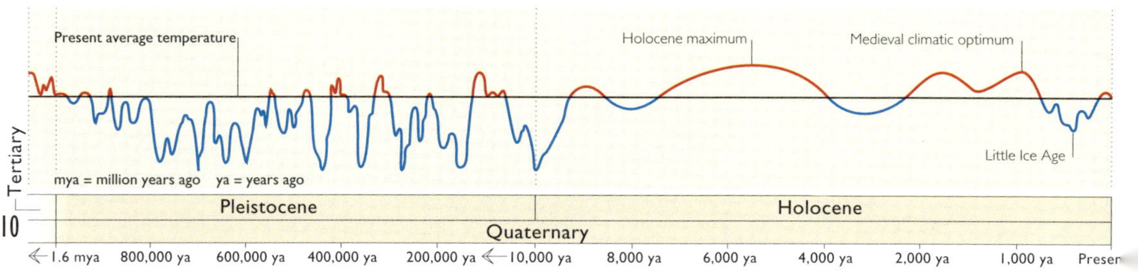

A PROLONGED DROUGHT *in North America's Southwest about 1280 led to the decline of Anasazi civilization and the abandonment of their cliff dwellings.*

the privations of Iceland in the following years are clearly linked to this poor weather.

In Europe, the cooling of the climate may have triggered population decline (which began even before the Black Death plague of the late fourteenth century). However, there were also a number of human factors involved. These included a rapid population growth, the extension of agriculture into less fertile land, and the occupation of vulnerable coastal sites.

THE GREENLAND COLONY

The demise of the Norse colony in Greenland is the only recorded example of a well-developed European society being completely extinguished. In AD 985, during a particularly warm period, an expedition led by Erik the Red from Iceland enabled some 300 to 400 colonists to establish two settlements on the west coast of Greenland. By the early twelfth century, there were more than 300 farms with about 5,000 people. Although life must have been hard, the colonists maintained herds of cattle, sheep, and goats; exploited the plentiful wildlife; and received supplies from Iceland and Scandinavia.

During the twelfth century, well ahead of any climate change in western Europe, the weather in Greenland cooled sharply. Although it warmed slightly in the thirteenth century, the following century was even colder. These changes caused increasing storminess and more extensive pack ice around Greenland. As a result, visits from Iceland became less frequent and were often separated by many years. The last contact was in 1410, and the settlement died out later that century. Archeological studies of the graves of settlers present a harrowing picture of malnutrition and debilitating diseases.

THE LITTLE ICE AGE

We are uncertain about the historic impact of the Little Ice Age—a cold period from about 1450 to 1850 that included many harsh winters. London's River Thames often froze over and was the site of frequent frost fairs (see p. 106). Series of cold, wet summers (such as in the 1590s, 1690s, and 1810s) were reflected in food crises throughout Europe and surges in Alpine glaciers.

The Little Ice Age is an elusive phenomenon and may not have been a single long-term episode at all. Recent studies suggest that it consisted of several cold intervals of up to 30 years each at the ends of the sixteenth and seventeenth centuries and from 1800 to 1820, interspersed by warmer periods.

RECENT DECADES

Since 1900, the global climate has warmed by about 1° F (0.5° C), concentrated in two periods—between about 1920 and 1940, and since the mid-1970s. It is still not clear whether this global warming is part of a natural cycle, or whether it can be attributed to human factors (see p. 126).

THE COLD WINTERS *of the Little Ice Age were recorded in Dutch and Flemish paintings such as* Hunters in the Snow—February, *by Pieter Brueghel (c.1525–69).*

Climate Change *and* *the* Environment

Life has persisted on Earth for more than three billion years, and has had to adapt to great variations in climate.

Over time, changes in the climate and the movement of the continents (see p. 112) have shaped how life has evolved on Earth.

According to the geological evidence, there have been five mass extinctions of flora and fauna. These may have been caused by great changes in climate, which may have resulted from meteorites hitting the Earth or immense volcanic eruptions. Each of these extinction events has heralded the end of a period: the Ordovician, 440 million years ago; the Devonian, 365 million years ago; the Permian, 245 million years ago; the Triassic, 210 million years ago; and the Cretaceous, 65 million years ago.

The consequences of these events are breathtaking: more than 99 percent of all species that have ever lived are now extinct. The total number of species, nevertheless, has increased. There are now an estimated 30 million species of organism, of which we have cataloged only about a million.

MIGRATIONS *During this century, as our climate warms, the armadillo (right) is expanding its range northward from the southern United States. The black rhinoceros (above), found in semi-arid climates, once flourished in the Sahara Desert, which was more fertile during a wetter period about 6,000 years ago.*

Evolution of Plants

A favorable climate allowed plants to spread onto land some 400 million years ago. During the past 80 million years, as the continents drifted toward their present configur-

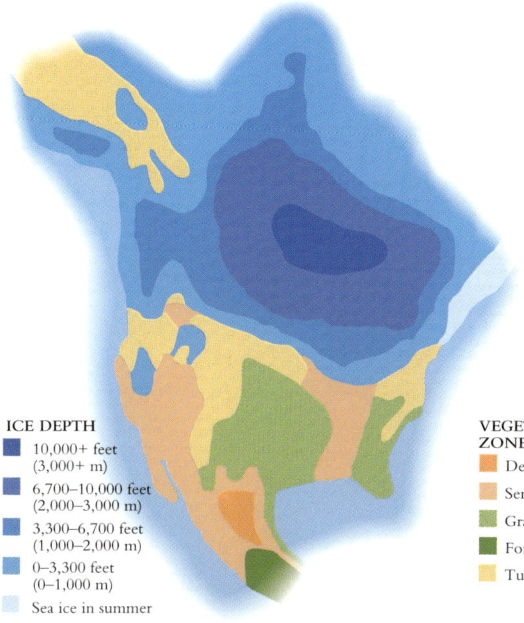

18,000 YEARS AGO, *at the height of the last ice age, much of North America was covered by glaciers.*

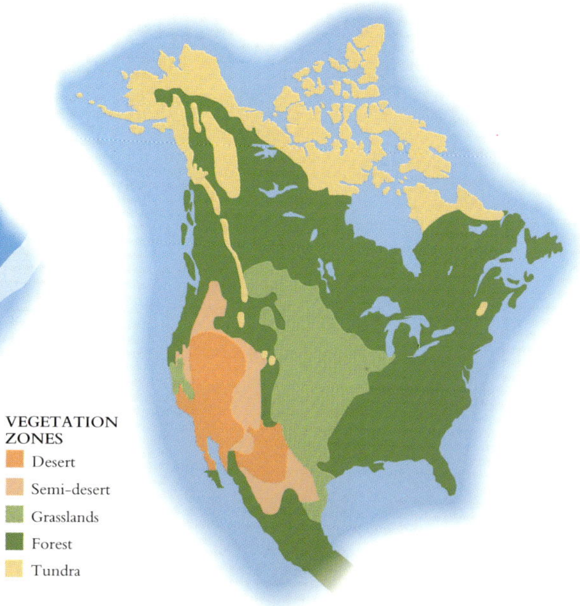

TODAY, *the areas of North America that were once under ice are covered in forest, grassland, and tundra vegetation.*

TRACES OF ICE *Glaciers gouged out U-shaped valleys, such as this one in Haines, Alaska (left). During the ice ages, the sweetgum tree (above) disappeared from Europe, but survived in North America because it could migrate south.*

ation, the relatively warm climate allowed flowering plants to diversify and become the most dominant group. Today, they number some 250,000 species.

By the Eocene epoch, some 50 million years ago, tropical plants were found as far north as western Europe and the northern United States, but the subsequent cooling trend altered this distribution.

The Last Ice Age

Most of the evidence of how life has adapted to climate change has been obscured by more recent events, with the natural world around us chiefly influenced by the dramatic oscillations of the past 1.6 million years.

The effects of the recent ice ages are clearly seen in the glaciated landscapes of the Northern Hemisphere, which include scoured U-shaped valleys and glacial boulders.

The distribution of flora and fauna contains more subtle clues about the events of the ice ages. In Europe, successive waves of ice eliminated species that are common in the eastern United States and China. These include kiwi fruit, tulip tree, and sweetgum—species unable to escape south across the east-west barrier created by the Alps and Pyrenees. In North America, the Rockies and Appalachians were no barrier to migration north and south. As the climate cooled, many species, such as Jack pine, retreated south. They returned north into present-day Canada as it warmed, although some remain in the United States in cool places such as mountaintops.

The rapid rise in sea levels about 12,000 years ago following the last ice age isolated some parts of the world from recolonization by plants and animals. When the British Isles were cut off from the rest of Europe some 7,000 years ago, some species were unable to spread north with the rising global temperatures.

LOUIS AGASSIZ AND THE ICE AGES

Theories of the ice ages are usually associated with the Swiss naturalist Louis Agassiz (1807–73). The possibility of ice sheets covering part of northern Europe had first been proposed in 1795 by Scotsman James Hutton (1726–97), the founder of scientific geology, but his ideas received scant attention.

In 1836, while on a field trip in the Jura Mountains, Agassiz became convinced that glaciers had transported blocks of granite at least 60 miles (100 km) from the Alps. In 1837, he first used the term ice age *(die Eiszeit)* and, in 1840, his proposals were published in a groundbreaking book, *Studies on Glaciers*. Initially, however, he was ridiculed by the scientific community.

Agassiz traveled to Scotland and Nova Scotia, where he saw more geological evidence of glaciation. In 1848, he joined the Harvard faculty and was active in many fields, notably marine science. He also continued glacial research in New England and the Great Lakes area. By then, it was clear that many features of the Northern Hemisphere could be explained only by ice ages, and Agassiz was vindicated, although the subject remained controversial in some quarters until the end of the century.

MEASURING CLIMATE CHANGE

Compiling a history of the Earth's climate demands a wide variety of sources and much detective work.

Instrumental records are the best source of information about climate change, although temperature and rainfall measurements are not always standard. For instance, early measurements of sea-surface temperature made from ships must be adjusted to account for the type of bucket used.

HISTORICAL RECORDS

Instrumental measurements were taken in only a small part of the globe and, at best, go back just 300 years. Ships' logs and weather diaries can fill some gaps; wine-harvest dates, cereal prices, and other records of weather-related events (such as the flowering of trees and shrubs and the freezing of lakes and canals) also help fill in the picture.

Where such records have not been kept, or to go back further in time, we rely on proxy data. Anything whose formation is both regular and affected by changes in the weather can provide such information. This includes tree rings, coral colonies, cores from ice sheets, and lake and ocean sediments.

TREE RINGS

As new cells are produced in a tree's trunk each spring and summer, a visible ring forms. Each ring represents a year in the life of the tree, and the ring's width indicates the tree's rate of growth in that year. The growth rate is usually determined by both temperature and rainfall. However, only where trees grow near the limit of their temperature or rainfall range can their rings be related mainly to one or other of the factors. Valuable sources of rainfall information include the firs along the northern edge of boreal forests in Canada and Russia, and the bristlecone pines high in California's White Mountains.

WEATHER LOGS *Beyond human records such as ships' logs (top), a history of climate is built up from clues in nature. Tree rings (below) provide an annual record, which, in the case of bristlecone pines (right), can extend back 4,000 years.*

The study of tree rings and climate (called dendroclimatology) was pioneered by Andrew E. Douglass in Arizona in 1904. This work helped to develop tree-ring records stretching back several thousand years by matching overlapping sets of narrow and broad rings from living trees and older timbers.

TROPICAL HISTORY

The instrumental records of the tropics are particularly limited, but the study of corals is helping to fill in the gaps. Corals live for centuries and grow by adding seasonal layers of calcium carbonate to their skeletons. Recent studies in Australia have shown that these layers can be linked to seasonal changes in rainfall and sea-surface temperatures.

FROZEN IN TIME *Antarctica's layers of ice extend our climate record back as far as 220,000 years ago, beyond the last two ice ages.*

Ice Cores

Ice cores are fruitful sources of climatic information. The thickness of annual ice layers records precipitation, and chemical analysis of the ice reveals the temperature at which the precipitation took place. Air bubbles trapped in the ice record the composition of the ancient atmosphere, the dust content is a measure of storminess, and the acidity signals major volcanic events around the world.

Cores drilled in the ice sheet of Greenland enable us to count individual annual snow layers for the past 15,000 years, and provide detailed insights into the climate for more than 100,000 years. They show, for example, that before the end of the last ice age, about 10,000 years ago, the climate was highly variable and average temperatures could rise or fall by 18° F (10° C) over the course of a few years. In Antarctica, the annual layers are too thin to count, but the thicker ice sheet extends the climate record back as far as 220,000 years ago.

Lake and Ocean Beds

To form a truly global picture and to be able to go back millions of years, climatologists study both the sediments on lake and ocean beds and the sedimentary rocks laid down throughout the Earth's history. They can sometimes identify annual layers, but usually rely on fossils in the rocks and sediments to discover how the climate shifted.

Seabeds contain the fossils of foraminifera, tiny ocean creatures that lived in the surface layer of the ocean. As each species of foraminifera could survive only in certain temperatures, their fossils help us identify shifts in ocean temperatures. Similarly, pollen from ancient plants that has collected in the ooze of lake bottoms records the climate changes that allowed different types of vegetation to flourish.

The Full Picture

Our attempts to build up a complete and accurate history of the Earth's climate are sometimes frustrated by unreliable evidence. Records kept by people are often incomplete because they tend to concentrate on extreme events rather than daily weather. Glaciers can wipe out evidence of previous advances, and ice sheets can distort over time, leading to the incorrect dating of the different layers. In ocean and lake beds, bottom-dwelling creatures might blur the record by stirring up sediment.

The climatologist's art is to tease out enough information from the variety of sources to draw reliable conclusions about past climate change.

CORAL COLONIES *growing near the coast are affected by river discharges, which, in turn, are linked to rainfall. By examining cross-sections of coral under ultraviolet light, scientists can identify past seasonal variations and build up a history of tropical climate.*

Weather Patterns

Meteorologists know that many of the weather's fluctuations are part of short-term patterns, but identifying long-term cycles is an elusive goal.

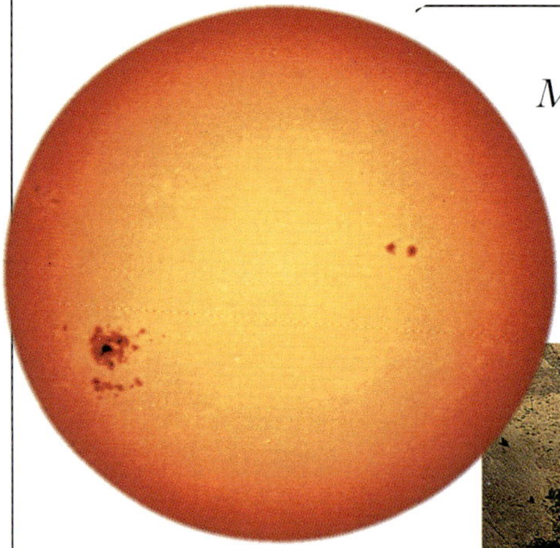

Although we're all interested in the weather, few of us think about long-term climate change until we experience unusual weather. However, it is important to understand that heat waves, cold snaps, and other spells may be part of a short-term weather cycle that is, in fact, still "normal" for the climate.

Everyday Weather

The climate of any given part of the world at any time of the year is defined by the averages for such factors as temperature, rainfall, wind speed and direction, and sunshine. Weather, on the other hand, is what is happening at any one particular time, and, more often than not, it is anything but average.

The Earth's middle to high latitudes experience frequent fluctuations that are quite normal for the climate. The prevailing westerly winds bring a succession of low-pressure systems, associated with strong winds and rain, that are interspersed with periods of high pressure, associated with more settled weather and clear skies. But even this standard pattern brings significant fluctuations in day-to-day weather: for instance, rain produced by low pressure may be concentrated in certain areas. And if the depressions are diverted from their standard tracks and cut off from the main flow—a process known as blocking—there may be spells of persistent weather. Memorable spells in recent years include the heat waves of 1976 in Western Europe and of 1995 in the United States.

At lower latitudes, the daily weather is more predictable, but there are still significant fluctuations in rainfall distribution. Many parts of the tropics and subtropics experience varying seasonal rainfall patterns, where years of drought are interspersed with wet years. For example, Zimbabwe and Mozambique experienced drought from

SUNSPOTS *(top) peak every 11 years, a cycle that may be linked to patterns of warm and cool weather, as well as extremes, from drought in Africa (above) to cold snaps in the United States (below).*

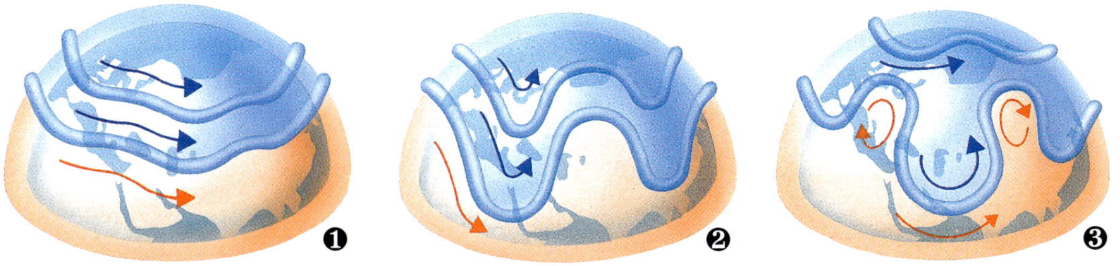

BLOCKING ❶ *The regular, meandering upper-air streams bring changeable weather to the middle latitudes.* ❷ *This flow is diverted from its normal westerly trajectory.* ❸ *If the pattern persists, high and low pressure systems may be cut off from the main flow and remain in one place for some time. Blocking brings spells of weather, such as heat waves or the especially cold winter of 1895 in England's Lake District (below).*

1982 to 1983, and again from 1991 to 1992, and a severe drought occurred over much of eastern Australia also during 1982 and 1983. In a sense, "normal" weather is drought plus flood divided by two.

The Search for Order

Variations over longer periods can also suggest a pattern. Weather statistics from around the world for the last 100 to 200 years reveal periods of "abnormal" weather that are a source of constant fascination to many meteorologists. However, the cycles come and go, and different parts of the world exhibit different cycles. Overall, the effect is a tantalizing glimpse of possible order within a sea of chaos.

One of the best examples of a weather cycle is the quasi-biennial oscillation (QBO) in the winds in the equatorial stratosphere. As the term implies, these high-level winds change from easterlies to westerlies every two years on average. The phenomenon has been recorded since the early 1950s, but it is still far from clear why it occurs and what effect, if any, it has on the general climate.

Faint cycles of between three and five years, found in many records, may be related to almost-regular fluctuations in the surface temperature of the equatorial Pacific Ocean. Linked to a warm ocean current known as El Niño that appears off the coast of South America (see p. 102), these fluctuations have become part of increasingly successful seasonal forecasts, but fall well short of a clear cycle that can be predicted years ahead.

A number of other theories has been proposed to explain longer-term weather patterns. Solar activity in the form of sunspots (dark areas that move across the surface of the Sun) follows a cycle, with a peak in the number of sunspots every 11 years or so. Magnetic-field fluctuations on Earth triggered by sunspot activity have an approximately 22-year cycle, while lunar-tide patterns recur almost every 19 years.

Numerous attempts have been made to link variations in the weather to a combination of these cycles. However, we are yet to identify a recognizable cycle that would allow us to predict weather patterns into the next century.

AVERAGE WEATHER STATISTICS FOR THE UNITED STATES *since 1905 reveal many periods of "abnormal" weather, as do records of the last 100 to 200 years from other parts of the world. However, an overall pattern is difficult to confirm.*

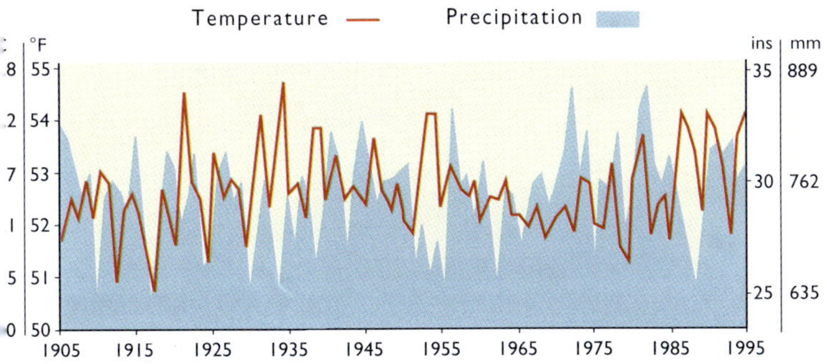

Tell me: where does this

Unexpected cold snap

Come from …

Weatherwise scarecrow?

Issa (1763–1827), Japanese poet

REASONS *for* CLIMATE CHANGE

Movements of air, water, landmasses, and the Earth itself may all contribute to long-term changes in our climate.

VOLCANIC ERUPTIONS *(left) force dust and gas into the atmosphere, and, in the past, may have led to ice ages. They are, however, much less frequent today and no longer have as great an impact on long-term climate.*

Long-term changes in the Earth's climate have been determined by a complex web of forces. Fluctuations in the patterns of atmospheric circulation, such as blocking (see p. 117), can produce spells of weather that persist for decades. Ocean circulation patterns, such as the great ocean conveyor belt (see box), can lead to climatic shifts that may last from years to millennia.

The much slower process of continental drift has also altered the climate. As the landmasses have either congregated closer to the equator or moved toward the poles, there have been entirely different climatic regimes. When there is plenty of land at high latitudes, a platform is created on which ice sheets can easily build up. The movement of land toward the poles also opens up huge tropical oceans that absorb heat, leading to global cooling.

The movement of the continents has also created major landforms, such as the Himalayas and the Tibetan Plateau, the Andes, and the Rocky Mountains. These, in turn, have had a profound effect on our present climate, accounting for much of its variability (see p. 36).

OUR PLACE IN SPACE

Extraterrestrial influences on our climate may include both long- and short-term variations in the amount of radiation emitted by the Sun, with short-term variations linked to sunspot activity (see p. 117).

The most important theory of long-term climate change, however, relates to variations in the Earth's orbit. Known as the Milankovitch theory, after the Yugoslav geophysicist who developed the idea in the

THE GREAT OCEAN CONVEYOR BELT

The "great ocean conveyor belt" may be the key to long-term changes in the surface temperatures of the oceans, which are an important factor in global climate (see p. 38). Cold, salty water, which sinks into the deep ocean in the North Atlantic, flows south and then east around southern Africa to resurface and be warmed in the Indian and North Pacific oceans. Surface currents carry warmer water back through the Pacific and South Atlantic. The round trip takes between 500 and 2,000 years. The latest studies suggest that the strength of this transport can easily change speed or direction. This could explain sudden climate changes, such as the Little Ice Age that affected Europe in the seventeenth century. It may also help us predict future changes.

There are no long-term observations of the conveyor belt, but measurements taken in the 1980s suggest that the formation of cold, deep ocean water in the Greenland Sea had declined by 80 percent since the 1970s. Similar changes may have caused sudden climatic shifts in the past. Moreover, recent changes in ocean-water temperature may have contributed to other climatic fluctuations, such as the sustained drought in the Sahel since the late 1960s, reduced hurricane activity in the Atlantic, and a rise in El Niño events in the tropical Pacific (see p. 102).

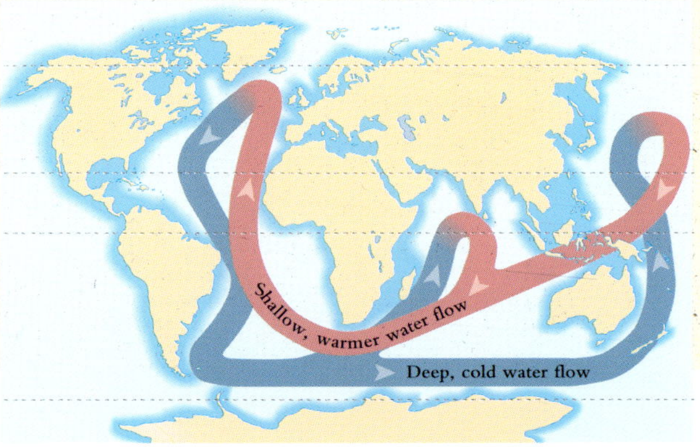

The Earth's axis **wobbles** like a gyroscope and traces a complete circle about every 23,000 years.

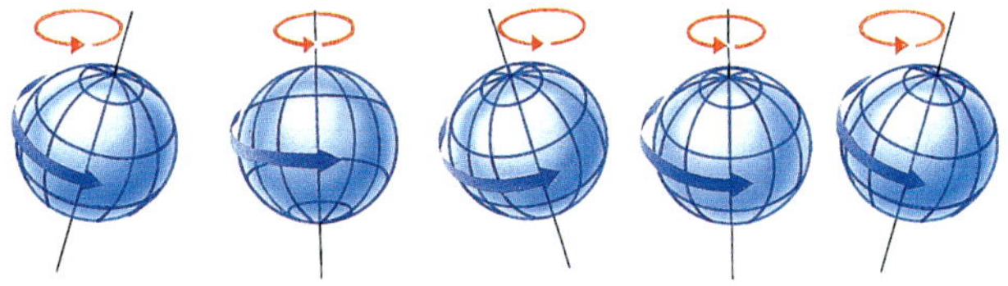

THE MILANKOVITCH THEORY

links three variations in the Earth's position relative to the Sun.

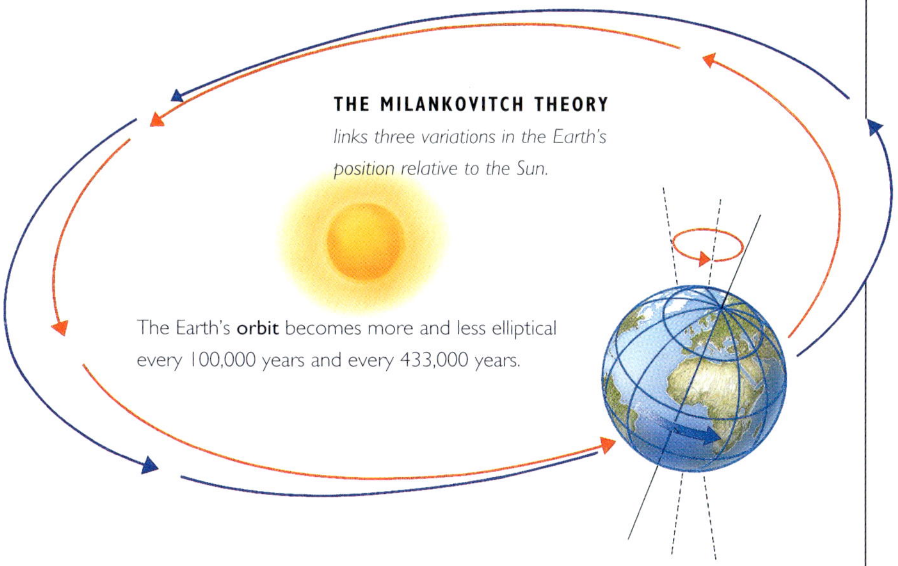

The Earth's **orbit** becomes more and less elliptical every 100,000 years and every 433,000 years.

The **tilt** of the Earth's axis fluctuates back and forth between 22 and 24.5 degrees every 41,000 years.

1930s, it links three periodic changes in the Earth's annual path around the Sun with the progress of the ice ages. These changes in the Earth's orbit alter the amount of sunlight falling at different latitudes: first, the Earth's axis wobbles like a gyroscope, tracing a complete circle every 19,000 to 23,000 years; second, at the same time, its tilt fluctuates between 22 and 24.5 degrees every 41,000 years; and finally, its orbit pulsates, becoming more and less elliptical every 100,000 and 433,000 years.

The Milankovitch theory is supported by studies of the amount of sunlight falling at high latitudes in the Northern Hemisphere, which has varied by up to 9 percent within each 100,000-year period. Further evidence is that the variations in the Earth's orbit closely match the waxing and waning of the ice ages. In analyses of Antarctic ice cores, the three shorter periods show up clearly over the last 250,000 years. The same cycles, as well as the 433,000-year one, appear in many geological records, such as the varying thickness of sedimentary layers, going back over millions of years.

GALACTIC THEORIES

Over the last 1,000 million years, ice ages have occurred roughly every 150 million years; only around the end of the Jurassic period is an icy epoch missing. This more or less regular behavior may be a coincidence resulting from the way the continents have drifted, or it may have an external cause. One possible explanation is that every 150 million years, as our solar system orbits our galaxy, it passes through dust lanes that border the galaxy's spiral arms. The dust reduces the sunlight that reaches the Earth, leading to an epoch when temperatures fall and ice sheets can form at high latitudes.

The above causes and others may all contribute to changes in our climate. Long-term patterns depend on the complex interaction of any number of factors, and we can only speculate as to what our future climate will be like.

THE MILKY WAY

Bordering the spiral arms of our galaxy are dust lanes that may be responsible for climate change. As our solar system passes through these lanes about every 150 million years, the Earth receives less sunlight.

The Human Impact

Most human activity has a barely noticeable effect on the weather, but where we alter the surface of the Earth substantially, the effects can be dramatic.

ACID RAIN *and other pollutants have killed these trees in Mynydd Dinas, Wales. The rain's acidity would have been increased by emissions from the nearby steel plants and oil refineries. Acid rain is also damaging our monuments and buildings (above right), from the Parthenon to the Statue of Liberty.*

There is still much debate about exactly what effect human activity will have on long-term climate, but we have clearly affected the weather on a local scale.

Urban Weather

The weather in large cities is appreciably different from the surrounding countryside. Buildings and roads absorb a lot of sunlight and store it efficiently, and human activities, such as industrial processes and air conditioning, create heat. Because of this, cities are warmer than their surroundings, especially on calm, clear nights, when the temperature in the center of the biggest cities can be as much as 18° F (10° C) higher than in the nearby countryside.

The extra heat causes air to rise more over cities, which, together with the smog and dust produced in urban areas, increases cloud formation and produces 5 to 10 percent more rainfall—notably, in the form of heavier thunderstorms in summer. The rapid runoff of heavy rain from asphalt and concrete surfaces increases the chance of flash flooding in urban areas.

Although skyscrapers can create wind tunnels, the effect of a large group of buildings is to form an uneven barrier that lowers the average wind speeds at ground level. These calmer conditions lead to the build-up of photochemical smog in summer and an increased probability of fog in winter. The smog combines with the higher temperatures to make heat waves more oppressive in cities.

However, cities do have their advantages. The higher temperatures in winter reduce heating bills, and the lower incidence of frosts means urban gardeners can grow more temperature-sensitive plants than can their country cousins.

Many features of our cities cannot easily be changed, but to improve the quality of urban air, we must reduce pollution. Maintaining open areas and planting more trees can help

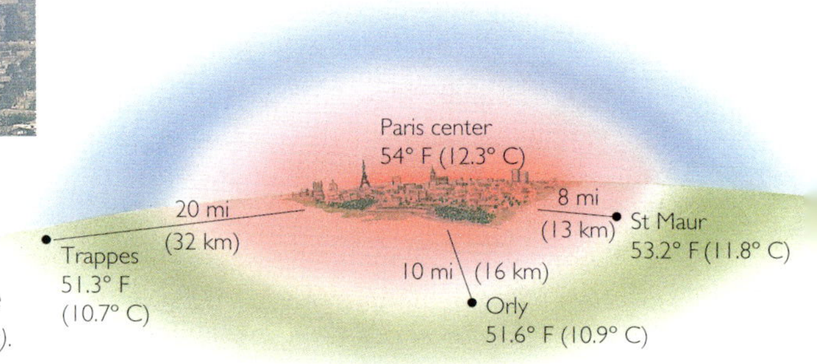

URBAN AREAS *affect the weather in many ways. Pollutants from industry and automobiles lead to thick photochemical smog, as occurs in Santiago, Chile (above). Human activities and buildings create heat islands in and around cities such as Paris (right).*

prevent flash flooding and other water-runoff problems. Furthermore, in summer, plants release moisture, which reduces temperatures, and trees provide valuable shade, which cuts the demand for air conditioning.

Changes in the Rain

Record-breaking floods on the Mississippi in 1993 and on the Rhine in 1995 have raised doubts about the benefits of conventional flood-control systems. Rather than building ever-higher levees, it may be better to permit controlled flooding of rivers. Because levee systems accelerate the speed at which water is drained away, they can make the effects of the heaviest rainfall worse (see p. 228).

Human activities have also affected the content of rainwater. Rain is naturally acidic, but in recent years, more than 10 times the normal level of acid has been routinely detected in rain over parts of Europe and North America. This acid rain contaminates water supplies, leaching vital nutrients from the soil, damaging forests and crops, and poisoning organisms throughout the food chain.

Acid rain is the direct result of burning fossil fuels. Cars and factories release sulfur and nitrogen oxides, which interact with sunlight to form sulfates and nitrates. When water vapor forms in clouds, it reacts with these, producing sulfuric and nitric acids, which fall to the ground as acid rain.

Deforestation

Deforestation certainly stirs strong emotions and can have devastating consequences for the environment, but its effect on global climate is unclear.

The destruction of tropical rain forests greatly affects the local climate, leading to higher daytime and lower nighttime temperatures. Studies suggest, however, that there is little impact on a global scale. The cleared ground surface reflects more sunlight out into space, which could have a cooling effect on global climate. At the same time, however, there are fewer trees pumping water vapor into the atmosphere, which, in turn, leads to less

We live inside a blue chamber, a bubble of air blown by ourselves.

The Lives of a Cell,
LEWIS THOMAS (1913–93),
American writer and academic

cloud and rainfall. The absence of cloud has a warming effect and compensates for the more reflective surface.

Oddly enough, the cutting down of the northern forests of Canada and Siberia may have a greater climatic impact because of the effect of trees on snow-covered areas. Snowy areas without any trees reflect more than two-thirds of the sunlight falling on them, whereas forested snowy areas reflect only about half of the Sun's rays. A computer model removing all forests beyond 45 degrees north latitude and replacing them with bare soil predicts significant cooling. At 60 degrees north in April—the thaw period when the absence of trees would have the greatest effect—the drop in temperature was estimated to be 22° F (12° C).

It has been argued that at least part of the global cooling during the last 5,000 years (see p. 109) can be laid at the door of deforestation.

THE ROLE OF FORESTS *in global climate is still unclear. The removal of the northern coniferous forests (below) may have a greater impact than the loss of tropical rain forests (above).*

Ozone Depletion

The appearance of the ozone hole over Antarctica during the 1980s is the most dramatic example of human activities altering the atmosphere.

Most of the ozone in the atmosphere is found in the stratosphere at altitudes between 9 and 25 miles (15 and 40 km). Ozone is a gas of three oxygen atoms that absorbs most of the harmful ultraviolet rays from the Sun and also prevents some heat loss from the Earth.

The discovery of an enlarged hole in the ozone layer has had a profound effect on scientific thinking about damage to our environment. It has also confirmed the fears of environmentalists who, since the early 1970s, have claimed that the ozone layer would be damaged by high-flying aircraft, the increased use of fertilizers, and, most of all, industrial processes that release chlorofluorocarbons (CFCs) into the atmosphere.

AEROSOL CANS *(above) used CFCs as a propellent, until recently. CFCs are also used in refrigeration and in foam plastic for fast-food containers and packaging.*

The Ozone Layer

The reason the ozone layer is so vulnerable is that there is so little of it. The total amount of ozone between us and space is equivalent to a layer encircling the Earth that is only about ⅛ inch (3 mm) thick, although this amount is, in fact, spread unevenly throughout the atmosphere. Most ozone is produced over the tropics, where solar radiation is strongest and most direct. Ozone is transported around the Earth by prevailing high-level winds. It may be depleted by increases in solar activity (see p. 116) and by the dust and gases released by major volcanic eruptions (see p. 270). Human activities can also disturb the fragile balance by producing long-lived pollutants that reach the upper levels of the atmosphere.

CFCs interfere with the normal process of ozone formation because they are capable of reaching the upper atmosphere, where they can form chlorine compounds that destroy ozone. This combination occurs over Antarctica at the end of the southern winter. Each October, an intensely cold vortex forms in the atmosphere over Antarctica, creating ice clouds and trapping CFCs. When the Sun returns after the Antarctic winter, the combination of sunlight, ice clouds, and CFCs forms an ozone-destroying mixture.

DEPLETION *of the ozone layer over Antarctica is likely to have a devastating effect on the flora and fauna, including these Dominican gulls and skuas (left).*

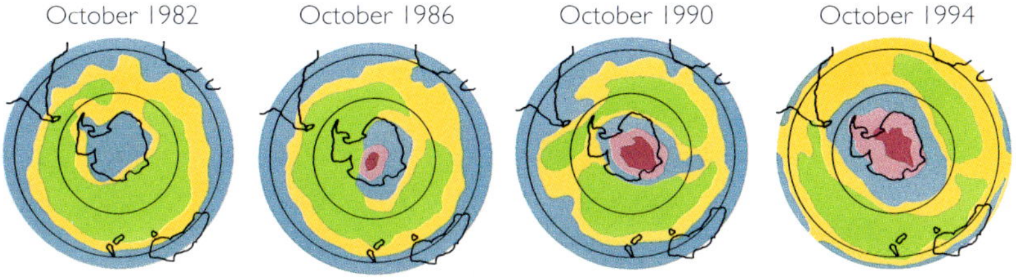

MONITORING CHANGE *Portable air-sampling units, such as this one in Hawaii (below), are used to monitor ozone depletion and other atmospheric changes.*

THE OZONE HOLE *over Antarctica, shown in dark red and pink (above), has grown since the early 1980s. The colors represent ozone concentration, from dark red for the lowest, through pink, blue, and yellow, to green for the highest.*

Antarctic Ozone

Each October during the 1980s, the amount of ozone over the South Pole plummeted, with lower levels being recorded almost every year. The scale and suddenness of the decline shocked the scientific world, and led to the Montreal Protocol in 1987 (and subsequent revisions in 1990 and 1992) to eliminate certain CFCs from industrial production. As a result of this rapid action, the global use of the most harmful CFCs fell by 40 percent within five years. Even so, ozone in the stratosphere over Antarctica had been almost completely destroyed by 1994, and it will be many decades before the CFCs already in the atmosphere are eliminated. Until then, ozone destruction will continue.

Changes in ozone levels in other parts of the world are much more difficult to predict. There has been a general decline of stratospheric ozone of a few percent in recent years. Some of this may have been caused by CFCs, but it was also affected by solar activity linked to sunspots, and the volcanic eruption of Mount Pinatubo in the Philippines in 1991.

There is some evidence of a hole forming in the ozone layer over the Arctic at the end of each winter, but it is far less pronounced than the Antarctic hole. Disturbing recent findings also suggest that an ozone hole is forming over Australia and New Zealand each winter, distinct from the ozone depletion occurring over Antarctica in spring.

THE ASH CLOUD *from the Philippines' Mount Pinatubo in 1991 (left) has exacerbated the decline of stratospheric ozone levels.*

We live in a period of atmospheric turmoil ... The sum ... is the drama of contrasts we call weather.

Knowing the Weather,
T. Morris Longstreth
(1886–1975), American writer

Low-level Ozone

Near ground level, the problems with ozone are entirely different. In major cities in summer, the mixture of pollutants, principally from internal combustion engines, is cooked up in bright sunshine to form photochemical smog, an important ingredient of which is ozone. As a result, lower-atmosphere ozone levels in populous areas have been rising. Although this ozone will help to screen out ultraviolet rays, it is not a real benefit as it damages vegetation and can be harmful to human health. It is also quickly absorbed and washed away by precipitation, so is much more variable than the stratospheric screen of ozone.

The Greenhouse Effect

Greenhouse gases make the Earth habitable, yet our activities may be increasing these gases and changing the climate.

CITY VEHICLES *release carbon dioxide into the atmosphere, thus contributing to the greenhouse effect and, possibly, global warming.*

If there were no atmosphere, the Earth's temperature would soar during the day and plummet at night, and the average temperature would be some 60° F (33° C) lower than the current value of around 59° F (15° C). This is because the atmosphere acts as a protective blanket warming the Earth, and maintains a constant balance between the amount of solar radiation absorbed and the amount of heat reflected back to space as infrared radiation. This balance sets the narrow temperature range in which life as we know it can exist on Earth.

Incoming Sunlight

The atmosphere cuts down the amount of sunlight reaching the Earth's surface. Clouds reflect about 30 percent of the sunlight falling on them and absorb some 15 percent of the sunlight passing through them. A clear, cloudless atmosphere absorbs only about 17 percent of the sunlight passing through it. Reflection from the surface of the Earth also plays a big part: fresh snow reflects up to 90 percent of the solar radiation falling on it, and desert sand about 30 percent. The oceans and tropical forests absorb up to 90 percent or more of all solar radiation.

Outgoing Heat

The Earth's surface also absorbs some sunlight and reradiates it as heat, or infrared radiation.

Most of the atmosphere is made up of nitrogen and oxygen (see p. 24), which absorb virtually no infrared radiation and allow it to escape to space. There are, however, "greenhouse" gases that absorb and reradiate the infrared radiation to the Earth. This warming process, which allows life to exist, is known as the greenhouse effect.

The principal greenhouse gases are water vapor, ozone, and carbon dioxide. Other trace gases such as methane, chlorofluorocarbons (CFCs), nitrous oxide, and sulfur dioxide also play a part.

The greenhouse analogy came about because these gases were likened to the glass of a greenhouse, which lets sunlight in but does not allow heat to escape. This helps to create a warm environment in which plant life can thrive.

The Enhanced Effect

Human activities, such as the burning of fossil fuels, are increasing the concentration of greenhouse gases, adding to the natural greenhouse effect.

The only way to calculate the impact of this build-up of greenhouse gases is by using computers to create global climate models (GCMs). If figures for the initial conditions of the atmosphere and oceans are fed into a model, it mimics global weather patterns.

RISING GASES *Burning fossil fuels releases carbon dioxide, and chemical fertilizers add nitrous oxide (above). The production of foam plastic uses CFCs (above right). When forests are cut down, we lose trees that absorb carbon dioxide (far right). Cattle release methane (right).*

CARBON DIOXIDE IN THE ATMOSPHERE

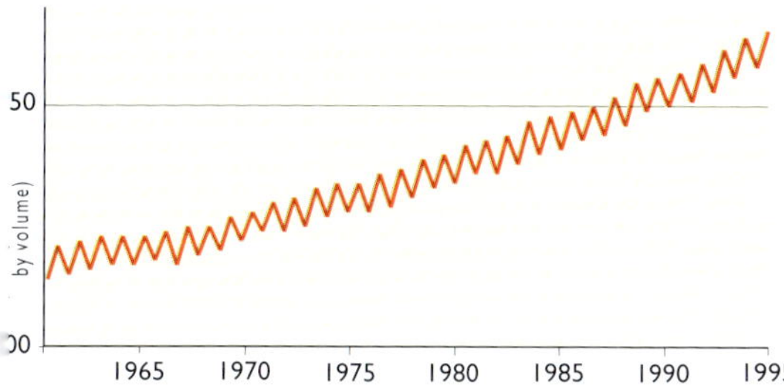

ACTUAL AND PREDICTED WARMING

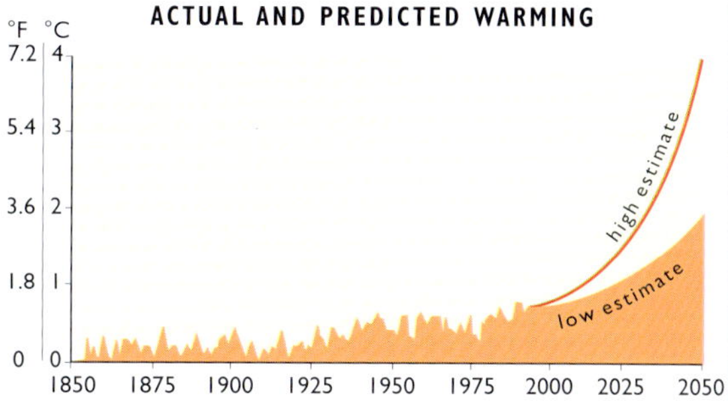

By running GCMs for the equivalent of many years, scientists can examine how the climate may respond to human activities. If we do nothing to curtail our production of carbon dioxide, it is expected to double by about 2060. Using GCMs, scientists predict this would lead to a 3 to 8° F (1.5 to 4.5° C) rise in average global temperatures, with polar regions warming by as much as 16° F (9° C) and the tropics by up to 5° F (3° C).

A Warmer World

Global warming would lead to increased temperatures and shifts in precipitation patterns, which, in turn, would have a serious impact on agriculture.

GCMs predict that the developed world is likely to benefit from increased warmth and rainfall in the middle latitudes. However, these shifts would bring drought and severe food shortages to the developing world in the tropics and subtropics. This conclusion contains the stark warning that global warming would increase the gulf between the developed and developing world.

THE GREENHOUSE EFFECT *is the warming process that allows life to exist on Earth.*

❶ Solar radiation enters the atmosphere, where a percentage is reflected or absorbed by clouds. When it reaches the Earth, some is reflected and the rest is absorbed and reradiated as heat, or infrared radiation.

❷ Infrared radiation (heat) leaves the Earth's surface and some passes through the atmosphere into space. The rest meets greenhouse gases, which absorb and reradiate it to Earth.

❸ Human activities, such as the burning of fossil fuels, are adding to the greenhouse effect by increasing the concentration of carbon dioxide, methane, CFCs, and nitrous oxide.

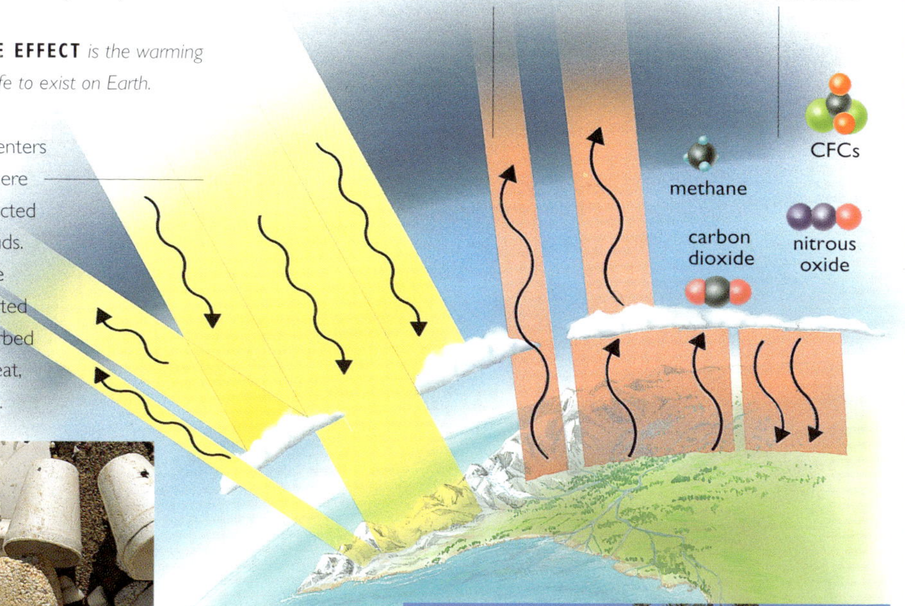

GLOBAL WARMING

We are still uncertain how much global warming is caused by natural processes and how much is caused by human activities.

The global climate has warmed by about 1° F (0.5° C) in the past hundred years. This matches the calculations of global climate models (GCMs) (see p. 125) that are based on an increase in carbon dioxide in the atmosphere, so there is growing conviction that human activities account for at least some of this warming.

Although GCMs are the only valid way we can predict climate change, there are some limitations to their usefulness. Long-term changes in the oceans (see p. 118) and their impact on global weather patterns are difficult to predict. We are also uncertain of the impact from human activities such as deforestation. In addition, there is the question of how much the current warming is due to the natural variability of the climate—this is far from clear from our present state of knowledge.

The biggest challenge in producing GCMs is estimating the precise effect of clouds.

OUR ENERGY NEEDS *have led to the burning of fossil fuels (right), which may be contributing to global warming. Public concern about our effect on the climate was expressed by a demonstration of pedal power in Berlin in 1995 (below).*

Clouds reflect sunlight, which has a cooling effect on the Earth, but they also absorb heat, which has a warming effect. We believe that they reflect more energy than they absorb, so, on balance, their effect appears to be cooling.

Our calculations are further complicated by the emissions of sulfur dioxide from power stations. The tiny, atmospheric sulfate particles increase the chances of cloud formation (see p. 42), and an increase in cloudiness is likely to reduce the sunlight reaching the Earth. Taking this into account, predictions of warming in parts of the Northern Hemisphere have been revised downward.

However, in 1994 scientists realized that clouds may absorb up to four times as much solar energy as had been assumed. This would have a substantial warming effect but is not included in the current predictions of global warming. Evidently, much more work

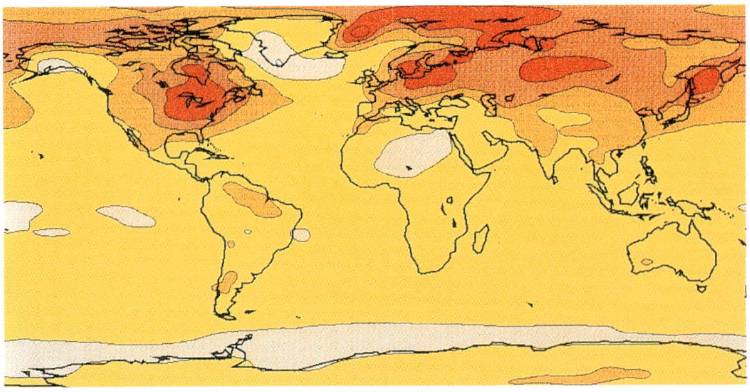

THIS CLIMATE MODEL *predicts the effect of doubling current levels of carbon dioxide in the atmosphere. Temperature increases are from pale yellow (0–4° F, or 0–2° C) to red (16–24° F, or 8–12° C). Dramatic warming occurs in northern Eurasia, North America, and the Arctic Circle.*

is needed on the properties of clouds before GCMs can handle them adequately.

Natural Variability

Before about 10,000 years ago, the climate was much more erratic (see p. 108). We do not know whether these changes were a feature of the last ice age and its aftermath, or whether they are normal climatic behavior and the stability of the last 10,000 years is unusual.

Air temperature is also only one factor in the complex web of climate. The rapid expansion of the world's deserts in the 1970s and the prolonged drought in the Sahel were originally seen as part of global warming. However, the situation was due to changes in rainfall patterns, which are now partly attributed to the El Niño effect (see p. 102). The drought and the expansion of the deserts probably had more to do with tropical sea-surface temperatures than warming of the atmosphere.

The warming observed in recent decades could partly be a natural fluctuation in either ocean circulation (see p. 118) or atmospheric patterns. Solar activity has also been held responsible for the warming, because the increases in global temperature closely parallel the 11-year sunspot cycle (see p. 117). The changes in solar energy are, however, far too small to explain the changes in temperature, unless they are amplified in some way by the Earth's atmosphere.

Preventing Warming

Doubts about the scale of future global warming do not mean that we should do nothing. At the very least, governments have to develop policies and set targets to delay the build-up of greenhouse gases. This resolve is reflected in the decision of the Rio Summit in 1992 to stabilize greenhouse gas emissions at 1990 levels by 2000. In 1995, the Climate Convention in Berlin agreed to work toward targets of lower emissions.

Translating these aims into effective action for both the developed and developing world poses many challenges. Taking action involves energy conservation, better low-pollution power plants and vehicles, improved industrial processes and public transport, and greater use of renewable energy sources. All of these changes are technically feasible but require political resolve on the part of governments and personal choices by individuals.

THE GAIA THEORY

In the 1970s, British physicist James Lovelock and American microbiologist Lynn Margulis stunned the scientific world with the hypothesis that the Earth itself could be seen as one giant, self-regulating organism. They called this organism Gaia, after the Greek goddess of the Earth. The Gaia theory holds that, while individual species have no knowledge of Gaia, they act collectively to try to ensure all aspects of the environment—the climate, the air, the sea, and the soil—are optimal for the survival of life in the broadest sense.

While this radical theory may leave many unconvinced, it does indicate a growing tendency to see the Earth as one entity, with all its parts connected. We are increasingly aware that human activity impinges on the Earth as a whole and, as may be the case with global warming, it can have unforeseen consequences.

RISING AIR TEMPERATURES *prompted the 1995 Climate Convention in Berlin (right). They were also blamed for the Sahel drought (above), although, in this case, ocean temperatures were the most significant factor.*

Chapter Six
Humankind *and* *the* Weather

Anything that lives where it would seem that nothing could live, enduring extremes of heat and cold, sunlight and storm, parching aridity and sudden cloudbursts ... testifies to the grandeur and heroism inherent in all forms of life. Including the human. Even in us.

Edward Abbey (1927–89),
American writer

WEATHER *and* HEALTH

The weather affects our health in many ways, and extreme conditions can put our lives at risk.

Although people have adapted to remarkably different climates throughout the world, the range of temperatures in which human life can exist comfortably and healthily is relatively narrow. We must maintain a core body temperature of about 98.6° F (37° C), and variations much above or below this can lead to illness: hypothermia and frostbite if we become too cold (see p. 134), or hyperthermia if we become too hot (see p. 132).

HOT AND COLD

In countries with marked variations in temperature, mortality rates are highest in winter, when circulatory, respiratory, and infectious diseases are prevalent. Cold weather cools the outer parts of the body, increases blood pressure, and places more stress on the heart. People with circulatory problems are hence at much greater risk and need to be careful not to overexert themselves.

Chilling of the body also reduces resistance to infections, ranging from the common cold to more serious diseases such as influenza. Furthermore, infectious diseases are easily spread in winter because people congregate indoors.

The low humidity of winter months can dry out human skin and lead to dermatitis. As cold weather restricts blood flow, it can also damage tissue and cause chilblains, especially on the fingers, toes, and ears.

Hot weather allows bacteria to thrive, so diseases spread by food and poor sanitation can be major killers in summer months. This was the case in the eastern United States until public health improvements at the end of the last century. Heat waves can still be devastating in developing nations.

The tropics and subtropics have a "malaria climate", with a wet season of heavy rainfall and high temperatures. Malaria, one of the world's greatest health problems, is spread by mosquitoes that need warm, wet conditions to complete their life cycle.

Warm weather can increase the incidence of hayfever and asthma attacks, because rising air currents carry pollen and other allergens long distances.

COLD CONDITIONS *reduce resistance to infection (top). A heat wave in Paris, 1895 (right). Sustained periods of hot weather can increase the mortality rate.*

THE WEATHER *has been blamed for everything from the misbehavior of schoolchildren (above) to suicide (left).*

Pain and Weather

Respiratory infections and muscular pains appear to be triggered by sudden changes in temperature and humidity that accompany both cold and warm fronts, especially when temperatures are low. Heart attacks, bleeding ulcers, and migraines have also been linked to abrupt weather changes.

Many people claim that they can forecast weather according to pains in their joints. Although we now have much more reliable forecasting tools, studies have shown that both the pains of arthritis and the "phantom pains" of amputees tend to coincide with the rapidly falling air pressure or increasing humidity of warm fronts (see p. 34).

We still do not know why these pains are sometimes brought on by the weather. Amputees' pains might occur because scarred skin expands and contracts in response to the weather at a different rate to the healthy skin around it. In the case of arthritis, perhaps a change in air pressure alters the pressure of the fluid in the joints, causing inflammation and pain.

In the Mood

The weather obviously affects our moods, and there is some evidence that conditions ranging from winter blues to suicidal states are linked to the weather. The connection is, however, still subject to conjecture, as these illnesses affect only a small percentage of the population, and no two people react in exactly the same way.

Windy days are often blamed for the misbehavior of children, while particular winds are associated with mental disturbances. The foehn wind, for example, descends from the Alps into northern Europe, bringing sudden changes in temperature, pressure, and relative humidity. These are thought to affect the mental health of some people and prompt symptoms such as headaches and insomnia. In fact, both the foehn and Italy's sirocco (a hot, humid wind) have been considered mitigating circumstances in criminal trials.

Heat waves seem to lead to violence. For example, the murder rate in New York jumped 75 percent during the heat wave of 1988, and rates of domestic violence consistently increase during hot periods. However, the link may be an indirect one: people tend to drink more alcohol during heat waves, and this is the more likely trigger for violence.

In North America, a long period indoors when snowbound is said to lead to "cabin fever", a condition causing irritability, depression, and anxiety. Recently, seasonal affective disorder (SAD)—a condition whose symptoms may include lethargy and depression—has been associated with the lack of sunshine in winter. It has been treated with some success by exposure to banks of bright lights for a short period each day.

IN HOT WEATHER, *crowds flock to the beach, but prolonged exposure to strong sunlight can lead to heatstroke.*

COPING *with* HEAT

Human beings first evolved in the hot, dry savannas of Africa, and all people are able to acclimatize to heat.

TROPICAL REGIONS *(left) do not usually get as hot as desert regions (above), but the humidity renders sweating less effective.*

We keep cool by sweating: our skin temperature is lowered as perspiration evaporates. Human skin is covered by more than three million sweat glands that can produce copious amounts of perspiration over the whole body, and our relative lack of body hair speeds up the evaporation. The price we pay is the risk of dehydration.

HOT AND DRY

In hot desert conditions, an adult walking briskly can lose more than 3½ pints (2 l) of water per hour. Without regular supplies of water in such weather, we rapidly dehydrate, which can lead to hyperthermia, a potentially fatal condition.

The relative humidity (see p. 40) determines how much this cooling process works: the higher the humidity, the less the cooling. Tropical regions do not generally get as hot as the more arid areas to the north and south, but they do experience an unceasing combination of high temperatures, humidity, and rainfall. In this weather, perspiration evaporates less rapidly, so sweating does not provide any significant relief.

The indigenous people of the hottest, driest parts of the world have evolved physically in a variety of ways that seem related to the climate. Low body mass, long limbs, and slim bodies make the sweating process more efficient. Long noses may help to humidify the air and reduce water loss from the lungs. Dark skin provides some protection from the strong sunlight.

However, medical tests have shown that all people can acclimatize to moderate heat quite quickly. People from

HEAT AND HUMIDITY

In hot weather, the combination of heat and humidity determines the level of human comfort. This comfort factor can be described in terms of either the apparent temperature scale—which combines actual temperature and relative humidity figures (see graph)—or the temperature humidity index (THI)—which is the mean of the dry-bulb and wet-bulb temperatures (see p. 99).

When the apparent temperature is above 90° F (32° C), about half the population feels hot and sticky, and by the time it reaches 105° F (41° C), most people are feeling uncomfortable. Summer heat waves with sustained periods above 105° F (41° C) are dangerous and can increase the mortality rate.

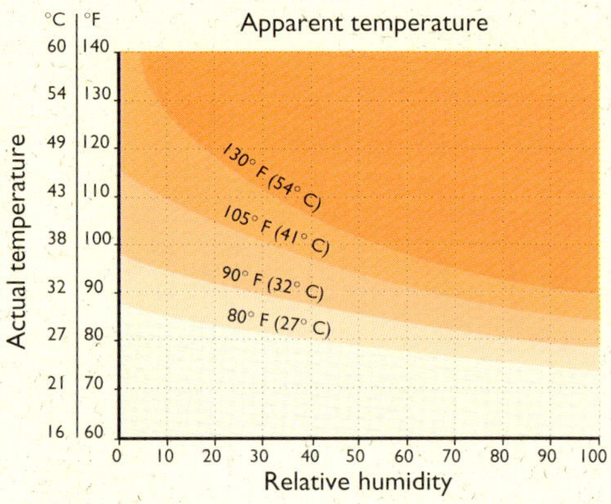

APPARENT TEMPERATURE *In the graph above, the curved line where the actual temperature and the relative humidity figures meet gives the apparent temperature figure.*

temperate climates who travel to hotter or more humid areas may initially feel most uncomfortable, but after 7 to 10 days, they sweat more (but lose less salt), their cardiovascular system is under less stress, and their body temperature rises less during exercise.

STILT HOUSES *in Queensland, Australia are designed to promote air circulation.*

Living in the Desert

The challenge of living permanently in the most arid parts of the world is all about one thing—water. In some regions, the scarcity of water and the sparseness of vegetation mean that native people have had to live a nomadic existence, moving from place to place in search of food and water for themselves and their animals. Where there is an adequate supply of water, a fertile island—an oasis—can be created in the desert and living conditions are good.

In industrialized countries, buildings are often air conditioned to control the heat of summer. However, as long as humidity is low and nights are cooler, low-technology solutions, such as appropriate clothing and building design, can provide relief from all but the worst extremes of heat.

Leaving your skin bare can make you hotter because it absorbs the Sun's radiation. There is also the risk of skin damage: a chastening statistic is that in Queensland, Australia, about 75 percent of people over the age of 65 have some form of skin cancer. The long, flowing robes traditionally worn by Arabs are an effective form of dress. They protect the skin from the damaging rays of the Sun, but also allow air to circulate so that perspiration can evaporate.

In buildings, thick adobe or brick walls provide insulation and shelter, and, being slow to heat up and cool down, smooth out the daily temperature fluctuations. For example, in the southwestern deserts of the United States, the widespread use in the past of heavy adobe walls and small windows produced dwellings that were pleasantly cool by day and relatively warm at night.

In other parts of the world, thick walls have been combined with artificial ventilation, and buildings are oriented to take advantage of prevailing winds or sea breezes. In many parts of the Middle East and the Indian subcontinent, wind towers or wind catchers on chimneys have been a form of ventilation for centuries. Chimneys themselves have been designed so that when they heat up in the midday sun, they create an additional through-draft even during calm conditions. In desert areas, such as Rajasthan, India, this type of ventilation is combined with moist reed mats that are hung in doorways. Any draft is thus cooled by evaporation and the result is a natural form of air conditioning.

WIND TOWERS *are used in the Middle East and Asia (left); the shaft extends down to living areas, allowing the breeze direct access. In hot, arid areas, loose clothing (above) helps air to circulate and perspiration to evaporate.*

Coping *with* Cold

Humans are not naturally adapted to extreme cold, and have had to create their own warm microclimates to survive in frosty regions.

Humans have a low tolerance of cold. Without protective clothing, we will start generating extra heat and shivering at temperatures as high as 68° F (20° C). Heat loss from the body can lead to hypothermia—a subnormal body temperature that causes progressive mental and physical collapse, and can be fatal.

Cold Protection

The best protection against cold is well-distributed subcutaneous fat. In this respect, women are generally better endowed, although they are often more sensitive to cold in their hands and feet. In addition, since heat production is in proportion to body mass, whereas heat loss is related to surface area, you are more likely to keep warm if you are heavy and stocky. For this reason, children are particularly vulnerable to hypothermia.

WINTER COLD *"Some enjoy it, others suffer it" (top); a Lapp child with reindeer in Norway (above); and Inuit building an igloo (left).*

Unlike our universal ability to adapt quickly to hot climates (see p. 132), we are generally unable to adapt to cold. The most effective response is to use clothing and shelter to create a warm, protected microclimate for ourselves.

However, our hands and feet do seem to have some adaptability, and tests have shown that people who work in cold, harsh conditions, such as deep-sea fishermen, are able to acclimatize slightly. With repeated exposure to cold, the blood vessels in the hands and feet develop the ability to dilate a little more effectively and maintain some warmth. However, this is a lengthy process, so when visiting cold climates, it is generally better to rely on clothing, equipment, and shelter for protection.

Living at the Poles

Polar regions have the most inhospitable weather on Earth, and provide some of the best examples of how humans can cope with extreme climate. The indigenous peoples of the circumpolar Arctic, such as the Inuit and the Saami, have evolved a way of life that enables them to survive the extremely cold winters in these regions. Although they have certain physiological advantages, such as additional fat and a stocky shape, it is their ability to produce a microclimate that is vital to their survival. For some 7,000 years, the combination of clothing, shelter, and diet has enabled these people to maintain a subsistence economy based on hunting and fishing.

Creating a Microclimate

The most important part of this way of life is efficient clothing. Clothes, mittens, and boots made out of animal pelts and furs can protect against temperatures as low as –76° F (–60° C).

The tent is a traditional form of shelter in the Arctic. By making such shelters out of caribou or reindeer hides, occupants can remain at a safe temperature on the most bitterly cold winter night with the aid of only a small fire. Another arctic shelter is the igloo built of snow blocks. Its design makes effective use of available materials, and produces the maximum living space with minimum surface area and heat loss.

Arctic Food

The Inuit eat large amounts of animal fats and no fresh vegetables. This diet is not generally regarded as healthy, but it provides plenty of energy to combat the cold. The Inuit do not have the same level of heart disease as people in industrialized countries who also eat a high level of animal fats. This may be because of the high proportion of fish oils in the Inuit diet, which apparently reduces the risk of cardiovascular problems, while helping to guard against extreme cold.

THE INUIT *of the Arctic make coverings for tents out of caribou hides.*

IT'S A CHILL WIND ...

The combination of wind and low temperatures can produce much greater heat loss from exposed flesh than the air temperature alone would suggest. The figure often used in cold countries to express this is the wind-chill temperature.

Wind-chill figures are based on experiments conducted in the Antarctic to estimate the risk of frostbite. When the wind-chill temperature is below –22° F (–30° C), there is a real risk of flesh freezing, and when it is below –58° F (–50° C), flesh will freeze in a minute or so.

In North America, where cold waves of arctic air can suddenly sweep southward, these figures are used for warning people of the dangers of going out. In warmer parts of the world, however, they can be misleading, especially when the temperature is above freezing. For instance, a combination of 40° F (4° C) and a wind speed of 30 miles per hour (48 kph) equates to a value of 12° F (–11° C). At this figure, it is highly unlikely that anything will freeze; indeed, climbers and skiers need to be on guard for although it may feel bitterly cold, snow will be thawing rapidly, and on steep, snowy slopes, the risk of avalanches may be extreme.

Wind speed mph	Actual temperature °F: 40	30	20	10	0
15	23	9	–5	–18	–31
20	19	4	–10	–24	–39
25	16	1	–15	–29	–44
30	12	–2	–18	–33	–49

Wind speed kph	Actual temperature °C: 4	–1	–7	–12	–18
24	–5	–13	–21	–28	–35
32	–7	–16	–23	–31	–39
40	–9	–17	–26	–34	–42
48	–11	–19	–28	–36	–45

WIND-CHILL TABLES *show the apparent temperature produced by the combination of actual temperature and wind speed.*

COPING *with* HIGH ALTITUDE

Adapting to altitude is not just about less oxygen; it also means lower temperatures, drier air, more sunlight, and sometimes more rain and snow.

Weatherwise, mountains tend to have more of everything except oxygen, which poses particular challenges for those who live at or travel to high altitudes. Anyone who moves quickly to an altitude of 10,000 feet (3,000 m) or more, whether going on a skiing holiday in the Rockies, trekking in Nepal, or taking a business trip to La Paz, will experience certain physiological effects when they first arrive.

Because of hypoxia, a lack of oxygen in the bloodstream, exertion will increase the heart rate and lead to shortness of breath, headaches, and possibly faintness. Eating too much can sometimes produce faintness and nausea.

HIGH ANXIETY

At an altitude of 16,500 feet (5,000 m), air pressure is half the value that it is at sea level and so there is only half as much oxygen available. Furthermore, the oxygen-carrying capacity of our blood is less efficient. At an altitude of 6,500 feet (2,000 m), it is reduced by 4 percent, but at 13,000 feet (4,000 m), it is reduced by 12 percent. The combination of these effects means that people breathe much more rapidly, increasing the amount of oxygen in their blood but reducing the carbon dioxide. This can lead to irregular breathing, headaches, and faintness—all of which can be dangerous for people with cardiovascular problems.

It is vital, therefore, for people visiting high altitudes to take things easy until they acclimatize. It also helps to drink plenty of water and to avoid alcohol, as dehydration resulting from the dry air can make the symptoms of hypoxia much worse. If symptoms persist, the only cure is descent. Altitude sickness can be fatal: fluid accumulates in the lungs and/or the brain, and can lead to drowning and/or brain damage.

Recent research indicates that although people initially seem to adapt to high altitude (see box), their work capacity is restricted, even after many months. The amount of oxygen-carrying cells in their blood increases over time, but

MOUNTAIN PEOPLES, *such as the nomads of Tibet (below), traditionally move their flocks to different altitudes according to the season. Edmund Hillary and Tenzing Norgay needed oxygen supplies to complete their ascent of Mount Everest in 1953 (above).*

THE HIGH LIFE *Mountain visitors, such as skiers in Canada (left), will never acclimatize as well as native highlanders, such as the Sherpas of Nepal (above).*

this is now seen as a sign of stress, and there is no evidence that it increases their aerobic capacity. When people return to low altitudes, however, their work capacity quickly reverts to normal.

Under the Sun

As we go higher in the mountains, the amount of damaging ultraviolet (UV) radiation in sunlight increases. The sunburning power increases by 4 percent every 1,000 feet (300 m) in altitude. At an altitude of 10,000 feet (3,000 m), there is 50 percent more UV radiation in sunlight than there is at sea level. This is because most of the gases and dust that absorb UV radiation occur below 10,000 feet (3,000 m). High reflection from snowy areas further increases exposure.

Peoples who have lived for many generations at high altitudes have skin types that tan easily and provide some protection from the powerful sunlight. For people holidaying in the mountains, though, especially those with fair skin, the message is clear: take extra protection against the dangers of sunburn. Even in late winter, skiing at high level can expose you to much more UV radiation than you would receive sitting on the beach at the height of summer. It is therefore essential to use high-factor sun cream and wear protective clothing and a hat when visiting the mountains.

... On crags piled who

knows how high,

A thousand different

grasses weep with dew ...

"Cold Mountain",
Han-Shan (eighth or ninth century), Chinese poet

ACCLIMATIZATION TO ALTITUDE

The differing adaptations of humans to the effects of altitude show how complex our physiological responses are. People born at high altitude who live in such regions all their lives have a much higher aerobic capacity than lowlanders. The Peruvian Indians of the Andes have unusually large chests and lungs that enable them to extract large amounts of oxygen from the air. The Sherpas of Nepal have a high oxygen-extracting ability that is related to the amount of hemoglobin, or oxygen-carrying molecules, in the blood.

Lowland adults who settle at high altitudes never become acclimatized, but young children taken to high altitudes develop adaptations similar to those of native highlanders. Highlanders do not show increased aerobic capacity when they descend to low altitudes.

CUZCO, *in the Peruvian Andes.*

It seems that many of the assumptions about the ability to acclimatize to altitude are unfounded. Initially, the response is good: after a few days, most people adapt, principally by better regulation of breathing. After this, however, there is little improvement and tests show that their work capacity remains well below sea-level values.

MODIFYING *the* WEATHER

For centuries, people have attempted to alter the weather, usually with very little success.

Prayers and ceremonies have been used by people throughout history in the hope of altering the weather. Among the best-known rituals is the rain dance of the Hopi Indians of North America.

In the nineteenth century, action of a more direct sort was taken by the mayors in Burgundy, France. In an attempt to prevent hail damaging their vineyards, they fired rockets into storm clouds. This futile practice was to continue well into the twentieth century.

At a more practical level, farmers have long grown shelterbelts of trees to reduce wind damage, and have irrigated crops with stored rainwater during dry spells.

Concerted efforts to alter the weather were made during the Second World War. Fog Investigation and Dispersal Operations (FIDO) involved the use of petrol burners to clear fog from English airfields. This method allowed some 2,500 aircraft to land during foggy weather, probably saving the lives of thousands of Allied airmen.

FIRING INTO CLOUDS *In the early 1900s, Clement Wragge tried using rainmaker guns (right) to break a drought in Australia. At about the same time in France, cannons were fired in attempts to prevent hail damaging crops (far right).*

CLOUD SEEDING

A major breakthrough in weather modification came in 1946, when American scientist Vincent Schaefer discovered that seeding clouds with tiny crystals of dry ice could cause precipitation. Dry ice has a very low temperature (-108° F [-78° C]), and the tiny crystals rapidly attract water droplets until they are large enough to fall as snow or rain. Conversely, seeding mature storms with dry ice can encourage the formation of smaller droplets, and so reduce the severity of the rainfall.

WIND FOR SALE *An etching from 1555 shows a sorcerer trying to sell a bag of wind tied up in three knots (left).*

Schaefer's colleague Bernard Vonnegut later found that silver iodide crystals formed ideal condensation nuclei (see p. 46) and so could also be used to seed clouds. A much cheaper option than dry ice, silver iodide is often used in today's seeding attempts.

ON A LARGE SCALE

The first serious attempt to modify major weather systems was made in 1947. Project Cirrus sought to reduce the winds in hurricanes by seeding.

THE RAINMAKER, *a film starring Burt Lancaster, tells the true story of Charles Hatfield. In 1915, during a drought in San Diego, USA, he agreed to bring rain for a fee of $10,000. His attempts were followed by devastating floods, and many lawsuits were filed against him.*

Unfortunately, the first hurricane that was seeded promptly altered course and caused widespread damage in Savannah, Georgia.

In the 1950s, despite various objections, commercial seeding operations were set up around the world to increase rainfall in times of drought. There were two main objections. Firstly, it was thought dangerous to tamper with complex systems that were little understood. Secondly, it was argued that the benefits of seeding were probably illusory, given that there was no way of telling what would have happened without it.

Weather modification received bad press in the 1970s. In 1972, a flash flood in Rapid City, South Dakota, which killed more than 200 people, was linked with local seeding activities. In the Vietnam War, suspicions about the United States using cloud seeding to flood the Ho Chi Minh Trail led to Senate Committee hearings.

Pressure grew for more rigorous statistical tests to be conducted on seeding throughout this period. The results of experiments were not encouraging. Only one experiment, to increase winter rainfall in the catchment area of the Sea of Galilee, showed positive results. There was an 18 percent increase in precipitation on seeded days, and the process did not appear to reduce precipitation in adjacent regions. For the rest, the conclusion was that despite 50 years' work on seeding, there had been no marked increases in precipitation.

When it comes to trying to alter severe storms, there are significant legal and political implications. The United States program Stormfury, for example, was aimed at reducing the intensity of hurricanes in the Caribbean during the 1960s and 1970s. However, much of the rain in northern Mexico comes from the remnants of these tropical storms, and it was claimed that the seeding had caused a severe drought there in 1980. The project was therefore shelved.

Weather modification is still attempted today, but on a limited scale and with less government funding than 20 or 30 years ago. Whether modern scientific techniques are any more effective than traditional approaches remains the subject of much debate.

CLOUD SEEDING *experiments with dry ice were conducted by Vincent Schaefer (left, front). Today, silver iodide crystals are released from generators mounted on light aircraft (below).*

HARNESSING *the* WEATHER

The weather has long been used to power machines and propel sailing ships. Today, wind, water, and sunlight are seen as renewable sources of energy for the future.

Since ancient times, the energy in rivers and streams and in the wind has been harnessed for human activities. There were some 6,000 water mills in England during the Middle Ages, used to grind grain and to power the bellows of forges. In Holland, windmills were used to drain and reclaim low-lying land.

Water still provides the most widely used form of renewable energy. Hydroelectric power uses water to drive turbines that produce electricity. The water is stored in dams, so the supply of power is not dependent on day-to-day weather. Prolonged periods of drought, however, can severely restrict the capacity of hydroelectric schemes to meet demand.

WIND POWER *Columbus relied on the wind to navigate the globe (top), and windmills have been used since the Middle Ages (right). Recently erected wind generators in California, USA (above).*

WIND AND WAVES

Windmills are very susceptible to weather fluctuations. For this reason, as electrification spread to rural areas, wind power was used only in the most remote sites. In recent years, however, the development of new technologies has made wind power the most competitive renewable energy source in windy parts of the world. Large numbers of big horizontal- and vertical-shaft wind generators are being tested in such places as Britain, California, and Denmark, and already are making an appreciable contribution to the electricity-supply networks. Other places with significant potential for wind power are the high plains of Oklahoma and Montana in the United States, and parts of India, Egypt, Argentina, and Chile.

Indirectly associated with wind power, the possibility of generating electricity from turbines driven by ocean waves is currently being studied in Britain, Norway, and Hawaii. Demonstration schemes show that although

HYDROELECTRIC SCHEMES *rely on the flooding of large areas to create dams, and can harm nearby ecosystems.*

this form of power generation is technically feasible, it is not yet economically viable. Even so, isolated island communities in windy regions may find wave power an economical, reliable source of electricity.

Solar Power

On clear days, the energy from the Sun can be used directly as heat or can be converted into electricity. In sunny parts of the world, solar-heat absorbers are widely used to supply domestic hot water. Buildings are now being designed to make use of sunlight for heat and light in winter, and to avoid overheating in summer. This reduces the amount of power used by electric or gas heaters and air conditioners.

The direct conversion of sunlight to electricity by solar panels has been used for many years to power space satellites and small pieces of electrical equipment, such as hand-held calculators. The cost of such panels has come down dramatically over the last 20 years or so, though they cannot yet compete economically with conventional power stations.

There have been many prototypes of solar-powered vehicles and solar-powered lightweight aircraft. However, at this stage, their purpose is to develop new technologies, rather than provide alternative forms of transport.

Ships and Planes

Sailing ships are no longer a viable form of commercial transport, but the weather still plays an important role in the routing of ships and aircraft.

Meteorological forecasts enable long-haul shipping to take routes that avoid head winds and heavy seas. This may mean the ships travel farther, but by exploiting favorable winds and light seas, they can cut the voyage time by as much as 10 percent. Similarly, airline pilots are advised of strong winds high in the atmosphere—known as jet streams (see p. 34)—which they can either ride or avoid to speed up the flight.

SOLAR ENERGY *powers this emergency freeway phone in Australia.*

The Impact on the Environment

Renewable sources of energy are less polluting than fossil-fuel power stations and they do not present us with the waste-management dilemmas of nuclear generators, but they are not entirely problem-free.

Hydroelectric schemes, in particular, are the subject of growing concern. The dams are created by flooding large areas, which alters the local ecosystem and displaces communities. In the long term, flooded areas are liable to silt up, depriving downstream areas of fertile sediment and affecting organisms throughout the food chain.

The large number of wind generators needed to make a significant contribution to electricity supplies can be unsightly—especially as the windiest sites tend to be in areas of outstanding natural beauty—and can present a hazard to birds, such as hawks and eagles, that hunt in these open areas. Wave-powered generators are an eyesore and might alter the ecology of nearby shorelines, while solar panels covering desert areas could affect the local climate.

The drawbacks of wind, wave, and solar power are relatively minor, though, compared with the predicted effects of global warming from the burning of fossil fuels, or the safety concerns about nuclear generators.

... you, pegged to a glacier's bitter stone
Outlast an Antarctic unending cyclone,
And watch, neither superb nor able, from below
A smashed world whirl away in stinging snow.

"On the Death of an Emperor Penguin in Regent's Park",
DAVID WRIGHT (b. 1920), South-African-born English poet and anthologist

Chapter Seven

Adapting *to* *the* Weather

EVOLUTION and CLIMATE

Plants and animals have evolved strategies that enable them to survive—and even thrive—in a range of climates.

WELL ADAPTED *Both the pincushion cactus (above) of desert biomes and the snowdrop (below) of snowy regions must cope with extreme conditions. The red-eyed treefrog (top) thrives in the constant humidity of tropical rain forests.*

The success of an organism depends upon the ability to reproduce young that grow up to produce more young. Only the young with the best survival traits for a particular niche or habitat will survive. These surviving young will pass on their adaptive traits to the next generation. This process is known as natural selection.

A region's climate may change over time, becoming hotter, drier, colder, or wetter. Natural selection will choose the organisms with traits best suited to the new climate. Over hundreds or even thousands of generations, continual small changes will eventually lead to new species that can thrive in the climate.

If climate or environmental change occurs too abruptly, species may not be able to adapt quickly enough and can become extinct. In recent times, humans have changed environments so rapidly that there has been an alarming rise in the extinction rate.

CLIMATES AND BIOMES

Biomes are large ecosystems of plants and animals that are adapted to a particular climate and soil regime. Climates and biomes change between the equator and the poles.

At the equator, there are warm tropical climates with high rainfall and little seasonal variation in temperature. These result in rain-forest biomes. Moving north and south from the equator, seasonal variations in temperature and rainfall lead to subtropical climates characterized by deciduous rain forest, or, in drier areas, savannas and scrub. At around 30 degrees north and south latitude, minimal rainfall creates great deserts.

Temperate climates, which have warm summers and cold winters, occur midway between the poles and the equator. The large transitional zone between temperate and arctic polar climates is known as northern temperate, and occurs from 50 to 66 degrees north latitude.

In polar climates, the Sun is always low in the sky and

VALLEY MICROCLIMATES *have lower nighttime temperatures and higher daytime temperatures, and also accumulate more water than the surrounding area. The vegetation is likely to be typical of higher latitudes.*

organisms must be able to withstand long periods of freezing temperatures with little or no light.

Both mountains and oceans have their own weather patterns. Mountains have higher precipitation and lower temperatures than the lowlands around them. Oceans give rise to coastal climates and semi-arid Mediterranean climates.

MICROCLIMATES

Microclimates are local variations within the normal climate of a region, and are caused by differences in topography, vegetation, and proximity to water or urban areas. Climate even varies at the same location, changing at different elevations. During the day, temperatures at ground level can be several degrees higher than at 6 feet (1.8 m) above the ground.

Microclimates influence the distribution of organisms in an area. For example, in the Northern Hemisphere, south-facing slopes receive more sun and less moisture than north-facing slopes. South-facing slopes therefore have shrubby vegetation, while north-facing slopes support lush forests. (The reverse holds true in the Southern Hemisphere.)

Temperatures inside a forest are lower on hot days and higher on cold evenings. Even trees planted around a house will create a microclimate by moderating temperatures and blocking wind and sunlight.

WARM AND COLD

Birds and mammals are termed endothermic (warm-blooded), meaning they can keep their body at a relatively constant, warm temperature by processing body fat and carbohydrates. All other groups of animals are called ectothermic, because their body temperature is determined by the temperature of the environment. The term cold-blooded is widely used to describe ectotherms, but it can be misleading, particularly in hot, arid climates. For example, a "cold-blooded" lizard basking in the hot sun can easily raise its body temperature to over 100° F (38° C), giving it "warm" blood—as long as it remains in the warm environment.

IN THE BLOOD *Reptiles such as the soft knob-tailed gecko (above) are ectotherms. Endotherms include the arctic fox (below).*

In cold climates, endothermic animals can remain active at night and during winter, but they must eat constantly. Other strategies adopted by endotherms include: minimizing heat loss with body insulation (hair, feathers, or fat); avoiding adverse temperatures by migration or burrowing; or reducing body temperature by hibernating.

In warm climates, organisms need to cool off. Animals' strategies include heat avoidance by retreating underground or to water; and evaporative cooling, usually through panting or sweating. Plants must adapt to heat and drought by using leaves and other structures to minimize heat stress and water loss.

TROPICAL CLIMATES

The climate of equatorial regions creates perfect conditions for tropical rain forests, which are teeming with plant and animal life.

Tropical climates occur mostly in the equatorial regions between the Tropic of Capricorn and the Tropic of Cancer. The constant light, warmth, and rainfall allow continuous, rapid plant growth, giving rise to luxuriant rain forests. The day length hardly changes throughout the year, and the weather stays very warm, even at night. Rainfall is high and often averages more than 100 inches (2,500 mm) per year. Even during the dry season, at least 4 inches (100 mm) of rain fall every month.

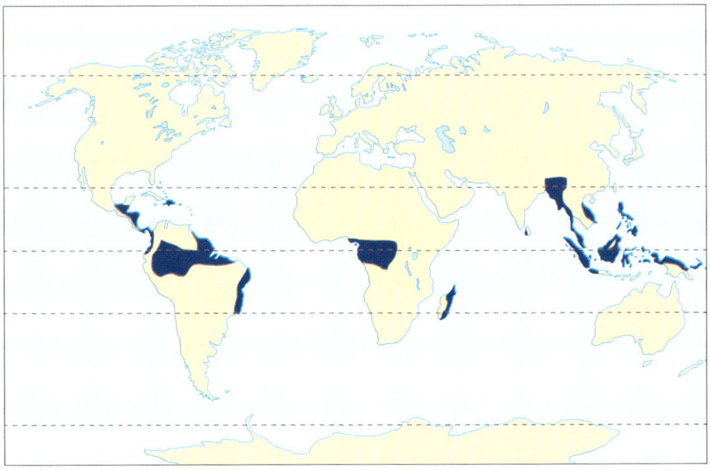

Tropical climates provide ideal conditions for life. Rain forests occupy only 7 percent of the world's landmass but contain 50 percent of the world's species—the greatest species diversity anywhere. So far, more than 1.5 million types of plant and animal have been discovered in these regions.

FOREST LAYERS

A tropical rain forest is divided from top to bottom into four distinct habitats: the emergent layer, the canopy, the understory, and the forest floor.

The emergent layer has the tallest trees, such as kapok and mahogany, which tower over the rest of the forest. These widely spaced trees have to tolerate intense sunlight, heat, and wind. The young leaves of some emergents are colored red by anthocyanin, a pigment that absorbs some of the Sun's damaging rays. The seeds of emergent trees are frequently dispersed by the wind.

The canopy consists of a solid cover of tree crowns, often forming several tiers of vegetation. The crowding causes knobby, club-shaped crowns and blocks most of the sunlight and wind. The large, broad leaves tend to be waxy with smooth edges, and have pointed tips that facilitate the rapid runoff of rainwater—damp leaves would become covered with fungi and algae, inhibiting photosynthesis. As canopy trees grow quickly toward the light, they tend to have long, straight trunks. Some have smooth bark to

RAIN FORESTS, *such as this one in Malaysia (left), provide abundant vertical niches for a diversity of fauna. Scarlet macaws (top) and chrysomelid beetles (above) inhabit South American forests.*

discourage parasitic plants and epiphytes (nonparasitic plants that grow on trees and obtain nutrients and moisture from the air). Since the canopy blocks wind, trees such as the chicle and cacao have evolved edible fruits with seeds that are dispersed by animals. Some trees have flowers on their trunks, making them easy for animal pollinators to find.

Beneath the canopy layer, the air is warm, still, and damp. Although about 80 percent of rainfall is absorbed by the overhead vegetation, the leaves are constantly releasing moisture, and the humidity is high (90 to 100 percent). The canopy also absorbs up to 99 percent of the sunlight, so it is dark inside the forest. The temperature remains stable throughout the year at about 80° F (27° C).

In the low light, understory plants are spaced apart, and have large leaves to maximize the light absorbed. Mimosa and other plants have a leaf joint, called a pulvinus, that rotates leaves toward the Sun.

Plants on the forest floor may have leaves arranged in rosettes so upper leaves do not shade lower ones. Ferns have divided fronds to maximize capture of sunlight. The forest floor has little soil because fallen leaves and branches rapidly decompose in the warm humidity and the released nutrients are quickly absorbed by other plants.

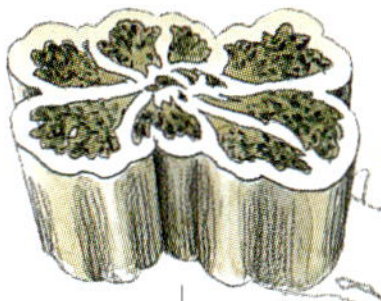

COMPETITION FOR SPACE *is intense in tropical forests, so epiphytes, such as bromeliads, and climbers, such as philodendron vines, grow on trees.*

IN THE RAIN FORESTS *of South America, the caiman (above) needs the warmth of a termite nest to incubate its eggs. This juvenile emerald tree boa (far left) will become green as it matures. The tree-dwelling lemur of Madagascar (left) has adapted to life in the canopy.*

TROPICAL FAUNA

A great diversity of animal life flourishes in the ideal climate of the tropics. The numerous plant species for animals to feed upon, along with the many vertical habitats of the rain forest, contribute to the abundant animal life. Many plants and animals have coevolved, forming mutually dependent relationships.

IN THE TREES

Many animals spend their lives high up in the trees. Primates such as lemurs, monkeys, and apes have many adaptations for living in trees, including rotating hands for grasping branches, and excellent stereoscopic vision for moving from tree to tree. Their teeth are adapted for an omnivorous diet consisting of a variety of plants, fruits, nuts, and small animals.

The rain forest supports the world's greatest diversity of birds. Toucans of Central and South America and hornbills of Africa and Asia, with their brilliantly colored, oversized bills, eat fruit from the tallest trees. Parrots and macaws eat fruit lower down in the canopy. Hummingbirds drink nectar and pollinate flowers.

Many kinds of fruit-eating bats come out at night, thus avoiding competition with diurnal birds for food. Moths are also common pollinators of night-blooming flowers.

Small lizards, such as anoles, are common tree-dwellers and important insect predators. Some use a sit-and-wait strategy to capture insects, thus keeping their body temperature only 2 to 3° F (1 to 2° C) above the air temperature. Larger lizards, such as iguanas, eat vegetation. They bask in the sun, raising their body temperature to speed up the difficult digestion of plant material.

WARM AND WET

Tropical mammals do not need to hibernate in the constantly warm climate. In fact, a hibernating rodent would be quickly devoured by army ants and fire ants, or consumed by snakes. Also, its fat reserves would be used up too quickly in warm weather.

Some mammals could not survive outside the warmth of the tropics because their slow metabolism generates very little heat. The slow-moving sloth hangs upside down in trees eating leaves and fruit. It has such an inactive lifestyle that algae grows in its fur, helping to conceal it from predators.

Many of the mammals that live on the forest floor cool off by sweating or panting. Some mammals, such as armadillos, also avoid heat by retreating into burrows.

The large, constricting snakes, such as pythons, anacondas, and boas, can

POISON-DART FROGS *Unlike other tropical frogs, which tend to be brown or green for camouflage, poison-dart frogs are brightly colored to warn predators that they are poisonous. The moist climate allows these frogs to carry developing tadpoles on their backs, so protecting the young from predators.*

A BROMELIAD NURSERY *Poison-dart frogs lay their eggs on leaves. As each tadpole hatches, a parent carries it to a water-filled cavity, such as a bromeliad reservoir. The female pygmy stawberry poison-dart frog returns regularly to feed the tadpole her infertile eggs. After about six weeks, the tadpole develops legs and leaves its watery home.*

attain their great size because the constant warmth and high humidity of the tropics allow them to eat throughout the year. These large snakes cannot digest food properly in cool temperatures and are likely to catch pneumonia in low humidity.

The constant humidity has enabled frogs to exploit habitats both on land and in trees. Frogs' eggs need to develop in water, while tadpoles must live in water because they breathe only through gills. As adults, frogs have lungs but they also breathe through their skin, which they must keep moist.

Termites and the Caiman

The abundance of wood, warmth, and humidity allows termites to flourish in rain forests. Their nests are found everywhere—as mounds on the forest floor or as nests on the trunks and branches of trees. Termites are able to digest the cellulose from wood with the aid of single-celled protozoa in their digestive tract.

The caiman, a relative of the crocodile, inhabits the cooler streams of rain forests, feeding upon terrestrial animals at night. This reptile needs a certain amount of warmth for its eggs to develop, but the shady forest floor is too cool. The caiman solves this dilemma by laying its eggs in the mound of a termite nest. The breakdown of organic matter in the nest releases heat, providing the caiman's eggs with the warmth they need to develop properly.

Going up that river was like going back to beginnings, when vegetation rioted ... and the trees were kings.

Heart of Darkness,
Joseph Conrad (1857–1924),
Polish-born English novelist

SUBTROPICAL CLIMATES

In some subtropical areas, the transition between tropical and temperate climates is marked by cloud forests and monsoon forests.

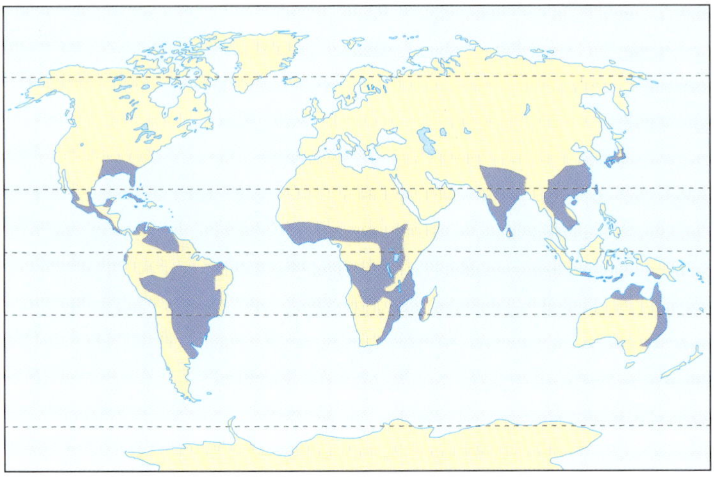

Moving north and south from the tropical regions near the equator, temperatures become lower and the rainfall becomes more seasonal.

Regions with a dry season followed by a period of heavy rain develop monsoon forests. Drier subtropical areas, as found in Africa, may contain savannas (see p. 156) or a shrubby form of vegetation known as thorn scrub.

In the mountains of tropical and subtropical regions, the air becomes cooled, water vapor condenses into mist, fog, or cloud, and the vegetation develops into cloud forests.

CLOUD FORESTS

Cloud forests grow on the mountain slopes of Africa, Central and South America, Indonesia, and New Guinea. The trees in these forests are constantly dripping, and, as in tropical forests, the branches are laden with mosses, orchids, and other epiphytes. However, the lower temperatures at high altitudes result in a mixture of temperate and tropical species, so that oaks may be interspersed with tree ferns, palms, and bamboo. Canopy cover is reduced and more light reaches the ground, causing greater shrub growth. In the cooler climate, leaf litter takes longer to decompose and builds up on the forest floor.

MONSOON FORESTS

Monsoon, or seasonal tropical, forests are found in southern Brazil, northern Australia, India, and parts of southern Asia. Because plant growth is greatly reduced during the dry season, which lasts five to seven months, the trees are shorter, more widely spaced, and have deeper roots. Most trees lose their leaves at the

CLOUDS AND RAIN *Cloud forests grow on mountains, as in the Bukit Barisan Range, Indonesia (left). Seasonal rainfall produces monsoon forests, as in Kakadu National Park, Australia (above).*

MEAL TIME *The collared anteater (left) eats ants in the wet season but switches to termites, which have a high water content, in the dry season. The emerald hummingbird (far left) drinks nectar from up to 2,000 flowers a day.*

start of the dry season and channel their energy into flower and fruit production. In turn, animal pollinators and seed-dispersers—such as birds, bats, wasps, and butterflies—become very active.

In Central and South America, the warm temperatures and abundant flowers create ideal conditions for the world's tiniest bird, the hummingbird. Because it is so small, this bird has a very high metabolism and must consume copious amounts of nectar. Its long bill and tongue enable it to drink from tubular flowers, which are not accessible to insects. To conserve energy on cool nights, the hummingbird becomes torpid, dropping its temperature from 100° F (38° C) to 64° F (18° C).

Animals living in monsoon forests have to adjust their behavior when their food supply changes with the wet and dry seasons. At the start of the dry season, some animals, such as lizards, cats, and monkeys, migrate to forests along riverbanks where the trees remain in leaf. Other animals, such as the collared anteater, change their diet. Rodents such as agoutis bury nuts in the ground and retrieve them when food is scarce. This behavior helps disperse the seeds of trees, as some nuts will be forgotten and can then germinate.

Birds that eat fruit and nuts, such as toucans and macaws, nest during the dry season when trees produce fruit and seeds. Insect-eating birds nest during the wet season. Insects flourish at this time because they can feed on the abundant young leaves.

THE COMMON IGUANA *lives in the subtropical forests of Central and South America. During the dry season, it minimizes water loss by excreting salt from nasal glands, instead of just in urine. Like most lizards, iguanas concentrate their urine, have waterproof keratin on their skin to reduce evaporation, and regulate their temperature by moving to shade or sun. Iguanas change color to a lighter green when they become hot.*

ARID CLIMATES

Many plants and animals have adapted to live with minimal rain, drying winds, and high temperatures.

Arid climates create desert, a biome where evaporation exceeds rainfall. Deserts receive, on average, less than 10 inches (250 mm) of rain per year, and in some years no rain at all.

Deserts experience huge daily temperature fluctuations. Because of the low humidity, over 90 percent of the Sun's radiation reaches the ground. Daytime summer air temperatures frequently exceed 100° F (38° C) and may reach 120° F (49° C). At sunset, heat is rapidly lost to the atmosphere because of the lack of insulating clouds. The nighttime temperature can drop over 80° F (44° C) from the daytime high, with winter temperatures in Asia's Gobi Desert plummeting to -5° F (-21° C). At high elevations, deserts may even receive snowfall during the winter months. Strong and frequent winds add to the drying conditions of the arid climate.

ARID LIFE *The lace cactus flower (top) and the Joshua tree (right) are found in North America's deserts. The quiver tree (below) grows in Namibia, Africa.*

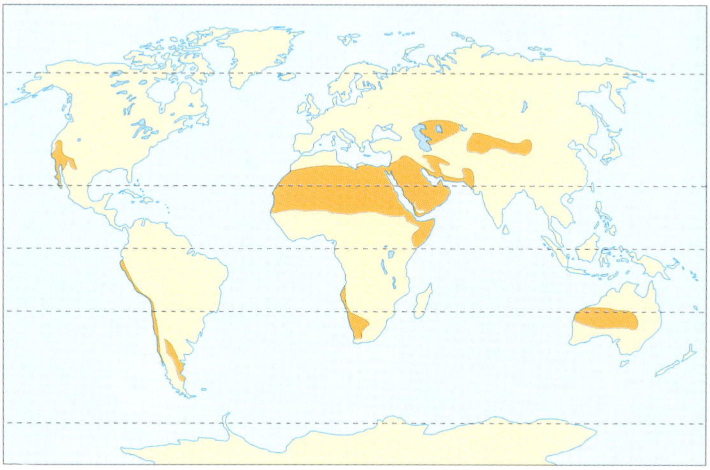

The largest deserts, such as the Sahara of Africa and the Great Sandy Desert of Australia, are located in two distinct belts around the Earth, near the Tropic of Cancer and the Tropic of Capricorn. In these regions of stable air masses and high air pressure, the air is constantly sinking, which causes warming, low humidity, and infrequent rain.

The other great deserts include the Mojave, Great Basin, and Sonoran deserts of North America and the Gobi Desert of Asia. These areas are dry because they are a great distance inland or they are on the lee (protected) side of a mountain range, and the clouds and humidity that come off the ocean rarely reach them (see p. 36).

THE CREOSOTE BUSH *of the American deserts endures severe water stress. A huge, spreading root system maximizes the rain absorbed, and also releases chemicals that stop competing plants from growing nearby. Older leaves, twigs, and branches fall off between rains, and flowers form only after at least 1 inch (25 mm) of rain.*

Thirsty Plants

Water makes up 80 to 95 percent of plant tissue, and plants need water to carry out life processes such as photosynthesis. In arid areas, however, plants must be able to cope with a lack of water, a situation often made worse by drying winds and high temperatures.

Succulents—plants that can store water in their fleshy stems—include agave; prickly pear and cholla of the cactus family of America; and some species in the euphorbia family of Africa. These species have evolved similar strategies. Their extensive surface root system quickly absorbs rain before it sinks through the porous soil, and they minimize water loss by having a thick, waxy outer layer and no leaves. Their accordion-pleated surfaces allow the plants to expand or contract as they absorb or lose water. Their many spines deter grazing animals.

Some desert plants, such as creosote, can survive without water. Compared with other plants, they have smaller cells in their leaves, resulting in smaller and thicker leaves, smaller stomata (leaf openings), and more hair (if present). In addition, these plants have a thicker waxy layer, a larger root system, and shorter shoots and branches.

Some plants, including the Washington palm, evade drought by growing close to a constant source of surface water, such as an oasis. Others plants, such as mesquite, send down very deep roots that are able to tap into a year-round supply of soil moisture.

Among those plants that become dormant between rains is the ocotillo. Once all moisture has evaporated from the soil, the plant drops its leaves and stops growing.

After rain, the desert floor may erupt with a profusion of wildflowers, such as desert primroses and monkeyflowers. These are ephemerals, annual plants that grow only after rain, completing their entire life cycle and setting seed in a few months. The seeds contain chemical inhibitors that need to be washed away by at least 1 inch (25 mm) of rain before the seeds will germinate. Temperature also plays a role: some species grow only after a warm summer rain, while others germinate only after a cold winter rain. Ephemerals have colorful and fragrant flowers in order to attract many pollinators in a short time.

EPHEMERAL BLOOMS *appear and spinifex grass flowers following rain in Australia's Simpson Desert.*

THE CAMEL *can lose up to 40 percent of its body weight to dehydration and then drink 30 gallons (136 l) of water in 10 minutes. It minimizes heat gain by storing fat in its hump instead of throughout its body. Unlike most mammals, the camel varies its body temperature, which can reach 104° F (40° C) before it begins to sweat.*

Desert Fauna

Animals survive the drought and heat of arid climates through adaptation or evasion. Some always become dormant during the dry season, timing the rearing of young to coincide with periods of rain, while others enter dormancy only when food and water are scarce. Animals that remain active year-round use strategies such as migrating to cooler areas or becoming active only at night, or have physiological adaptations that allow them to cope with high temperatures and limited water.

The maximum body temperature that animals are able to tolerate is 113 to 122° F (45 to 50° C). In most climates, the best method of cooling is evaporation of water from the body surface by sweating or panting. However, this strategy is not effective when water is in short supply, and desert animals have developed other cooling methods.

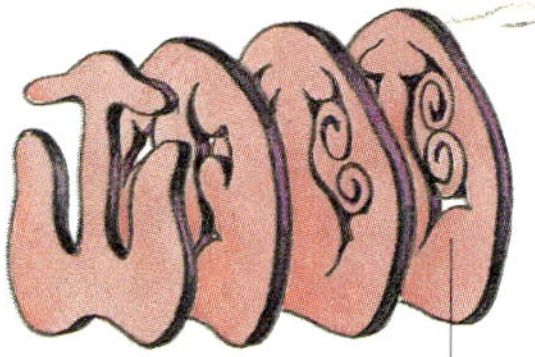

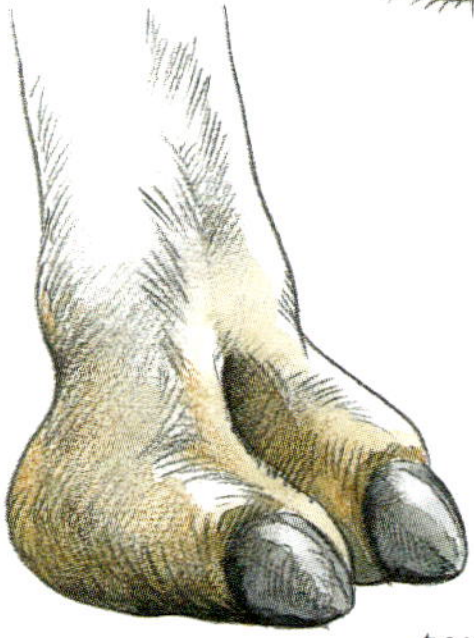

KANGAROO RATS *of the American deserts can go their entire life without drinking a drop of water.*

In Cold Blood

The body temperature of ectothermic (cold-blooded) creatures fluctuates with the outside temperature and is regulated by moving into or out of the sun. Snakes and tortoises retreat to burrows, while scorpions are nocturnal.

The metabolic rate of ectotherms varies with the temperature and can be very low. This allows them to eat about one-seventh less than a mammal of the same size does, an important advantage in the desert where food is often scarce. Reptiles lose very little moisture through their

HOT AND DRY IN AFRICA *The angulate tortoise (left) stores water in its urinary bladder. The thick-tailed scorpion (below) can tolerate a body temperature of 122° F (50° C).*

One look and you know that simply to survive is a great triumph …

LESLIE MARMON SILKO (b. 1948), Pueblo writer

scale-covered skin and concentrate their uric acid with their feces into a semi-solid whitish mass.

Some desert invertebrates, such as the scorpion and the black stink beetle, can tolerate more body heat than most other animals. Invertebrates have a waxy covering that inhibits moisture loss. A few also trap an insulating layer of pollen or air next to their body: blister beetles put pollen on their wings, while darkling beetles trap air. Aphids and weevils stay cool by drinking plant juices; other insects fly to cool places or go underground.

WARMING UP

Birds and mammals are endothermic (warm-blooded) and cope with arid environments in a variety of ways.

Desert birds cool off by soaring. Doves and quails concentrate their waste and urine. Many birds can survive a loss of 40 to 50 percent of their body weight in water, compared to a human capacity of only 12 percent.

When the temperature is too severe—hot or cold—insect-eating bats and most desert rodents will go into a dormant state, known as hibernation in cold weather and estivation in hot weather.

Kangaroo rats obtain all the moisture they need from seeds and other food. Their urine is concentrated so that only two-thirds is expelled as water and they consume their fecal matter to reclaim some water and vitamins. Their nasal passages efficiently reclaim water vapor from exhaled air. Like many desert animals, kangaroo rats live in burrows to stay cool during the hot days and warm on cool nights. In burrows, animals do not need to resort to evaporative cooling to lose heat and the relative humidity is much higher than outside.

Desert rabbits will shelter in the shade of a bush and wiggle their long ears. The ears radiate excess heat and present a large surface area that can be cooled by breezes.

The two-humped camel of the Gobi Desert and the one-humped camel of Arabia and the Sahara are well adapted to desert life. They survive by drawing on the energy stored as fat in their humps and on the water stored in their stomachs. This allows the camels to live many days without drinking anything.

LIZARDS, *such as the collared lizard, lie flat against warm rocks when cold. When too warm, they raise themselves off the ground by stretching their legs.*

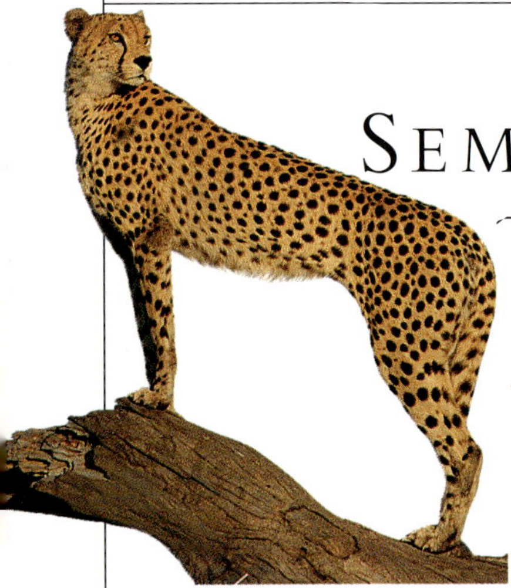

Semi-arid Climates

Displays of seasonal flowers and visits by grazing animals bring variety to semi-arid regions.

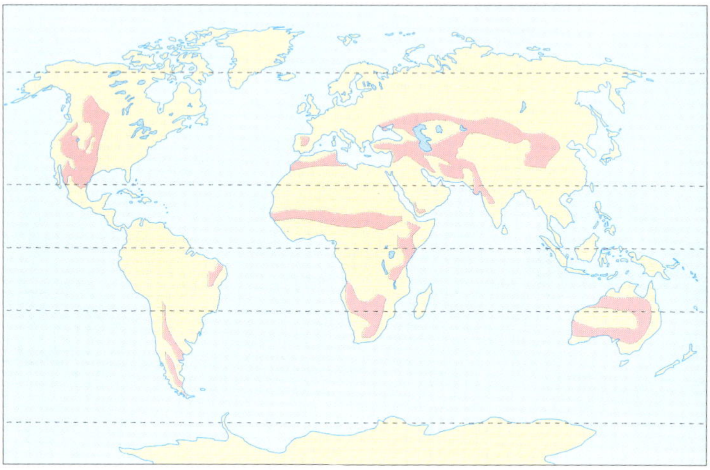

The semi-arid climate is characterized by low rainfall of about 10 to 30 inches (250 to 760 mm) a year, too little water to support a forest and too much to create a desert. Instead, semi-arid regions are dominated by large expanses of grassland, which are called prairie in the central United States, steppe in southern Russia, pampas in South America, and veldt in Africa. Savannas are grasslands with enough moisture to support sparse tree growth. Grasslands experience severe droughts periodically and are also very windy because they are flat and exposed.

The Grasses

Sweeping seasonal fires and grazing by large herd animals help clear the thick build-up of thatch that would otherwise choke the grasses. They also remove competing woody plants that would otherwise begin to colonize and dominate the grassland.

The largest percentage of a grassland's total living mass—its root system—lies hidden underground. The grasses' deep, thick roots store nutrients, collect the limited moisture from the soil, and resist being pulled out by grazing animals. Grasses can recover quickly from disturbances such as fire and grazing because their

IN SPRING, *prairies are carpeted with colorful wildflowers, such as bluebonnet, paintbrush, and flax (left).*

AMPHIBIANS, *such as the spadefoot toad (right), need water to begin their life cycles, but can still survive in semi-arid climates.*

THE CHEETAH *of the African savanna (left), the fastest land mammal, hunts medium-sized prey, such as gazelles and impala. African zebras (right) spend 75 percent of their time grazing.*

creeping root system quickly colonizes barren areas and allows them to sprout new shoots in numerous locations.

Grassland Animals

The dominant animals of grasslands are the large herds of hooved grazers, such as zebra, gazelles, and wildebeest in Africa; camels in Asia; bison, pronghorns, and antelopes in the United States; and several species of deer. Grazing animals' migratory behavior helps prevent an area from being overgrazed. Animals gather in lower grasslands near water in the dry season, and head for fresh pasture on higher ground when the rain returns.

Like most animals, grazers give birth in spring, when young have the best chance of survival: water is still ample and there is plenty of fresh, new growth for the nursing mother to feed upon.

The gazelle, a small, swift antelope, is adapted to heat and drought in a number of ways: its coat reflects sun and heat, it cools by nasal panting, and its hairy nasal passages recover most of the moisture it exhales. Gazelles tend to feed at night or in the morning when plants have the highest water content.

Because grasslands have few trees, most birds nest on the ground and are not strong fliers. The ostrich of African grasslands is the world's largest bird, and has long, powerful legs to outrun predators. It can go without water for days and withstand air temperatures of 133° F (56° C), an important adaptation in a treeless habitat. The South American rhea and the Australian emu fill the corresponding niches in their grassland ecologies.

Much of the animal life in grasslands is found underground. Rodents—such as the prairie dog of America, the gerbil of Africa, the hamster of Eurasia, and the widespread vole—retreat underground to escape predators and the summer heat. Hamsters and prairie dogs hibernate during cold winters. Gerbils live in drier areas and have water-conserving adaptations such as deriving most of their water from their food and producing concentrated urine.

THE SPADEFOOT TOAD *survives the long, dry months by digging deep into the earth using its rear feet and then encasing its body in a waterproof, cocoon-like coating.*

The toads come out of dormancy when they hear the sound of thunder. They quickly lay and fertilize their eggs in temporary ponds, and the eggs hatch in two days. Ten days later, the tadpoles have already lost their tails, developed legs, and left the pond—a much quicker development than that of other toads.

MEDITERRANEAN CLIMATES

Mild winters and springs, and dry, warm summers mean that the ability to survive on little water is a feature of successful species.

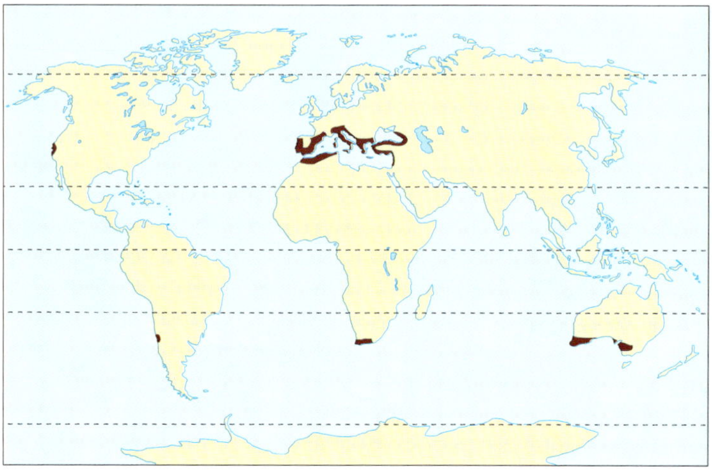

Mediterranean climates mostly occur on coasts that lie between subtropical and temperate zones, and exhibit extremes of both these zones.

Because of the proximity to oceans, winters are mild and moist, and summers sunny and dry. Typically, there is little or no rainfall for four to six months, and most plant growth occurs during spring, after the winter rains. These weather patterns often lead to summer wildfires.

SHRUBS AND ANNUALS

Dry summers, wet winters, and soils low in nutrients favor the growth of shrubs and dwarf trees. Mediterranean shrublands are called maquis in regions around the Mediterranean Sea, chaparral in California, mattoral in central Chile, fynbos in South Africa's Cape region, and mallee in southern Australia.

Shrubs have a number of advantages over trees and herbs. Their woody, compact shape reduces evaporation and heat stress during summer, while their extensive roots gather the limited moisture and nutrients from the soil. Being evergreen, shrubs can retain nutrients from season to season. Many Australian shrubs are able to recycle nutrients such as phosphorus from leaf litter.

Mediterranean shrubs reduce water loss with leaves that are small, thick, and waxy. Most are also stiff-leaved, as leaves bent by the wind would lose more moisture. Some shrubs, such as the kermes oak, have leaves with spiny margins that help dissipate heat and discourage grazing by animals. Others, such as rosemary, roll their leaves inward, trapping air inside to reduce evaporation. Many shrubs are covered with hairs, which trap air, help shade the leaf, and dissipate heat. The olive has light gray foliage that reflects sunlight.

Some shrubs, such as sage-brush and eucalyptus, have

MAQUIS AND CHAPARRAL *Wild olives grow in the maquis of Tunisia (above). Chaparral fires sweep through California in summer (right). The omnivorous scrub jay (top) can obtain all its water from its food.*

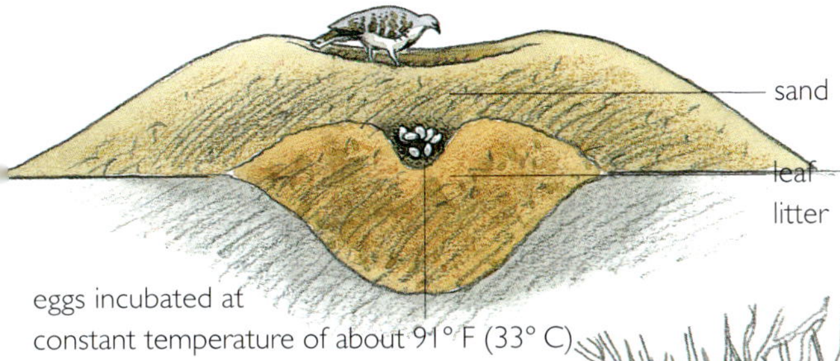

THE MALLEE FOWL *of Australia builds the base of its nest from leaf litter, which rots and generates heat. The male uses its sensitive mouth and tongue to check nest temperature, then adds or removes a layer of sand to make the nest warmer or cooler.*

aromatic foliage full of oils that deter animals and inhibit other plants from growing nearby, thereby limiting competition for water and nutrients. These oils add explosive fuel to wildfires.

After a fire, many Mediterranean shrubs resprout from root crowns. The seeds of some species need fire to germinate and may lie dormant for years until the next fire.

These climates also suit annual plants, such as poppies. After the soil dries, annuals die and release seeds that remain dormant during the summer months and germinate when the winter rains return. Other plants, such as wild onions, die back to an underground bulb or tuber that stores nutrients through summer, allowing for rapid growth the following spring.

Mammals and Birds

Animals survive fire by taking flight to other habitats, or by retreating to underground burrows. Animal populations flourish on the lush plant growth that follows every fire.

Native mammals include deer, rabbits, and numerous rodents. On hot summer days, rodents, such as voles and ground squirrels, retreat to burrows to cool and conserve moisture. The nocturnal Mitchell's hopping mouse of the Australian mallee can survive without drinking by eating porcupine-grass seeds.

In southern Europe and California, many of the native animals have been replaced by large domestic grazers such as cows, sheep, and goats. This heavy grazing pressure has altered the originally shrubby landscape into one favoring annual grasses, such as wild oats and brome.

Much of the birdlife is migratory, visiting mainly in spring and fall. Resident birds tend to have short wings and long tails, an aid to maneuvering around shrubs. Most shrubland birds are opportunistic feeders, eating insects in spring, seeds in summer, and berries in fall.

EUROPEAN RABBITS *eat their own waste in order to extract the maximum amount of nourishment from their food.*

TEMPERATE CLIMATES

Temperate climates have four distinct seasons, so flora and fauna must adapt to variations in both weather and food supply.

Temperate climates are found in the eastern United States, Europe, and eastern Asia. They have warm, humid summers and cold winters with snowfalls. Annual precipitation ranges from 20 to 60 inches (500 to 1,500 mm). During the winter months, freezing temperatures can lock up moisture as snow and ice, creating a virtual drought.

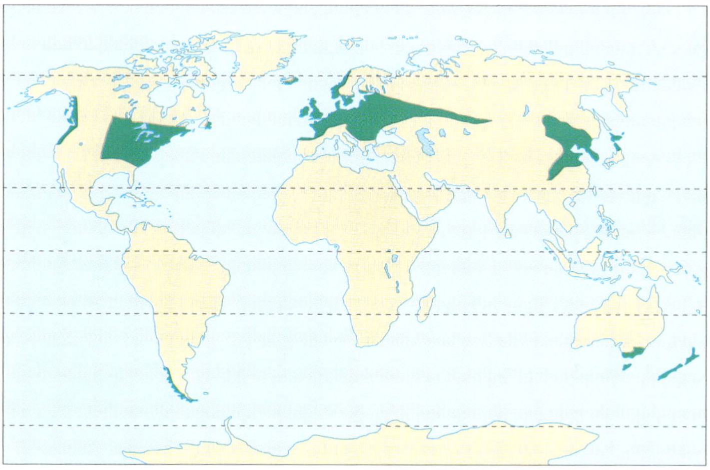

DECIDUOUS FOREST

The dominant vegetation in temperate climates is dense, deciduous forest. Common tree species include maple and birch in moist areas, and oak and hickory in drier areas.

The thick tree canopy blocks 90 percent of sunlight during summer, so the lower leaves on trees are broad and thin to maximize the amount of light captured. Leaves growing on the tops of the trees receive more intense sunlight than the lower leaves, and are smaller, thicker, and shinier—characteristics that help limit water loss.

Broadleaf trees lose a lot of water through their thin leaves, and need a constant source of soil moisture to avoid drought stress. In temperate climates, however, soil moisture may freeze during winter. Deciduous trees solve this problem by losing their leaves in fall. As the day length shortens and temperatures drop, minerals are transferred from the leaf to the stem, and the leaf begins to die. The green chlorophyll breaks down, and pigments become visible, creating stunning displays of color.

IN THE WOODS, *maple trees display fall colors, while the white-tailed deer is starting to grow its thick winter coat (left). The European fire salamander breathes partly through its skin and must keep it moist year-round (top).*

In fall, the leaf litter from deciduous trees helps create rich soils—which, in turn, encourage a great diversity of invertebrates and fungi. In spring, before the trees are covered with leaves, flowering plants grace the forest floor.

TEMPERATE ANIMALS

Animals in temperate climates need to survive cold winters and seasonal variations in their food supply. Some vary their diet from insects to seeds to berries, or between different kinds of leaf.

Many birds and mammals deal with winter by migrating to warmer climates. Most amphibians and reptiles, as well as some large mammals, hibernate or become dormant. Some animals, including tree squirrels, remain active all winter, relying on stored

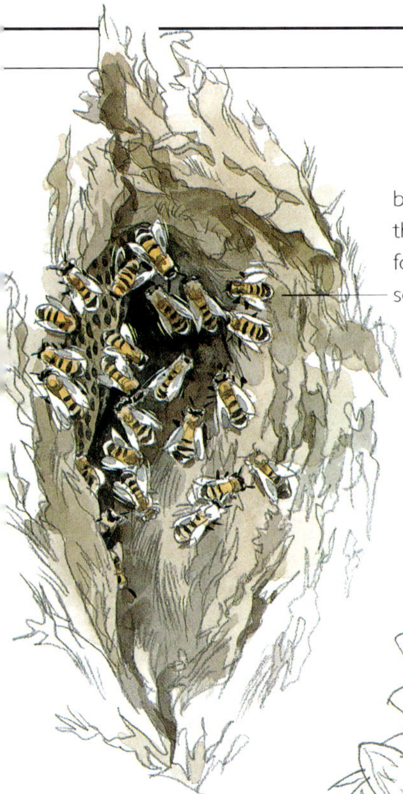

HONEY BEES *can cope with both warm and cool weather. They create evaporative cooling by putting water onto combs and fanning it with their wings. They warm up by exercising their flight muscles and basking in sunshine.*

food. Most adult insects lay eggs in summer or fall and then die. Their eggs hatch the following spring.

Turtles have an omnivorous diet, allowing them to eat a variety of plant and animal foods. Most turtles hibernate by burrowing into areas that are unlikely to freeze, such as the mud at the bottom of a pond. While hibernating underwater, turtles lower their metabolic rate, which reduces their oxygen needs. Many turtles absorb small amounts of oxygen directly from the water through thin skin protuberances called papillae.

Salamanders reach their greatest numbers and diversity in the temperate forests of Europe and the eastern United States. If a salamander is too dry, hot, or cold, it will retreat underground until it reaches moist soil of an appropriate temperature. During the cold winter, salamanders hibernate beneath the soil, emerging after the spring thaw to lay eggs in the abundant water.

Every spring, migratory birds leave the tropics and fly tremendous distances to nest in temperate forests. Birds risk these long journeys because of the plentiful food supplies offered by temperate zones in the spring and summer months.

WINTER TO SPRING *The painted turtle (right) produces an antifreeze that prevents ice forming inside its organs during winter. Birds, such as yellow warblers (left), that nest in temperate regions raise more young than they would in the tropics.*

Northern Temperate Climates

Despite often freezing conditions and a short growing season, a variety of life thrives in the northern forests and bogs.

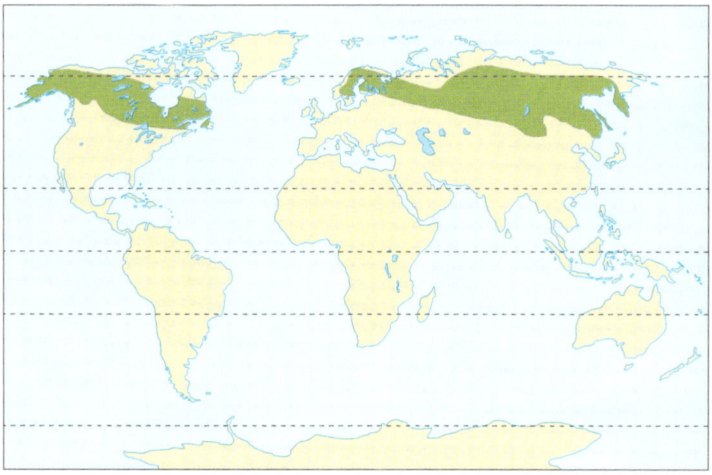

The northern temperate regions of Siberia and Canada contain the world's largest tracts of coniferous forests. This forest biome, called taiga or boreal, is a transitional zone between the arctic tundra to the north and the temperate deciduous forest to the south.

For up to nine months of the year, northern temperate regions have very cold, harsh weather. During much of this time, water exists only as snow or ice and little moisture is available to plants. The soils become waterlogged during the spring thaw and often stay soggy for a long time because of the permafrost, an underlying layer of permanently frozen soil. The slow decay of plants in the cold climate leads to extremely acidic soils, which trap nutrients needed for new plant growth.

Northern Plants

In summer, there is enough light and warmth for tree growth, but the winters are too harsh for most broadleaf trees to grow successfully. Although there are some broadleaf species such as birches and alders, the forests are dominated by spruce trees and other conifers such as firs and pines. Coniferous trees have a conical shape and needle-like leaves that help them survive the cold, snowy winters (see p. 164).

CARNIVORES AND CONIFERS *Plants such as the sundew (top) overcome the lack of nitrogen in boggy soils by trapping and digesting insects. Spruce trees (left) and other conifers are shaped so that snow slides off their branches.*

Northern temperate regions support ferns, mosses, and members of the heath family, as well as trees. Large expanses of boggy ground are carpeted with a dense, thick growth of sphagnum moss, which can survive freezing temperatures, immersion in standing water, and complete drying out. The stems and leaves contain large, hollow cells, so the moss is not damaged by expanding ice crystals and can also hold 20 times its weight in water during the spring thaw and rains. Dead sphagnum moss becomes peat, which can form layers several feet thick in northern bogs. Other bog plants include sedges, pondweeds, waterlilies, and algae.

The most common flowering plants are members of the heath family, such as blueberry, bog rosemary, and azalea. The roots of heath plants are able to absorb the

SNOWSHOE HARES *have huge feet covered with long, stiff hairs and spread-out toes, allowing them to travel easily over soft snow. Their thick fur changes from brown in the summer to white in the winter for year-round camouflage. They survive the winter by eating bark from trees.*

scant nutrients that bind with the acidic soils. Their leaves are small, thick, and succulent—all adaptations for the dry conditions caused by moisture being frozen in the soil during winter and absorbed by conifers in summer. Many species have waxy or hairy leaves to protect them from the drying winds. With the exception of blueberries, heath plants are evergreen and quickly begin photosynthesizing at the start of the short growing season.

THE BROWN BEAR *(left) has an omnivorous diet. It puts on large stores of fat in summer and hibernates in winter.*

Moose to Mosquito

Animal life is not abundant in northern temperate forests because of the cold winters and the low diversity of food sources. Most birds migrate to warmer areas for the winter months. Because large size limits heat loss, the mammals found in this habitat tend to be much larger than their relatives in warmer climates. Large animals living in this climate include moose, deer, elk, caribous, wolverines (of the weasel family), beavers, and brown bears. Smaller mammals include snowshoe hares, weasels, and squirrels.

The animals in northern temperate climates have to be resourceful in finding food. Moose, for example, frequently browse on aquatic plants, such as sedges, pondweeds, and waterlilies—a niche not grazed by other large mammals. The pine grosbeak eats many types of seed, and will eat fruit and insects in times of shortage.

The northern bogs are infamous for their swarms of mosquitoes. Mosquito larvae thrive in the shallow swamps, which freeze in winter and so contain fewer aquatic predators, such as amphibians and fish. With the first frosts, the adult mosquitoes die, but their eggs will survive winter in the frozen water until the next spring thaw.

Mountain Climates

Each of the three mountain zones—montane, alpine, and subalpine—has distinct flora and fauna, as well as species that migrate between the zones.

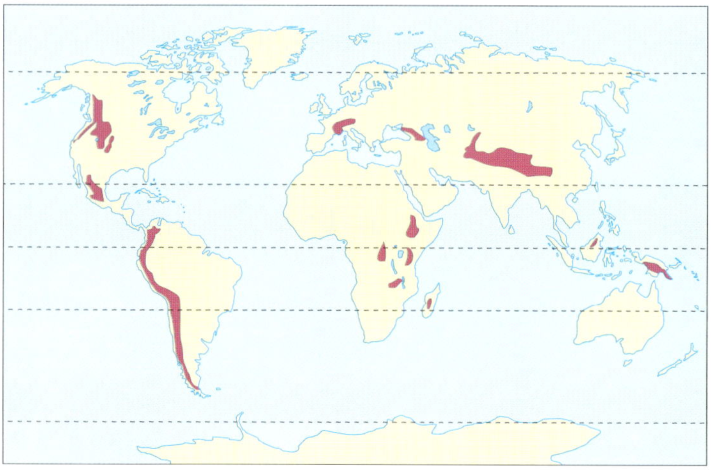

Mountains have a major influence on climates because they intercept and alter moving air masses, thereby affecting temperature, humidity, and precipitation levels.

Air cools while ascending a mountain slope, and its capacity to hold water decreases, causing an increase in rain and snow (see p. 36). The thin, dry air loses heat rapidly, and after sunset, temperatures plummet.

Mountain climates also tend to be particularly windy. Wind gathers speed at higher altitudes and as it travels over mountain ridges and through canyons (see p. 32).

Flora and fauna must therefore adapt to low temperatures and high winds. Furthermore, as altitude increases, air pressure falls and less oxygen is available. At an elevation of 10,000 feet (3,000 m), an animal needs to inhale one-third more air to obtain the same amount of oxygen as at sea level.

THE NORWAY SPRUCE, *like other conifers, is shaped so that snow slides off it easily. As it is evergreen, it can begin photosynthesizing straight after the spring thaw. Its thick bark contains chemicals that protect it from summer fires.*

YEAR-ROUND RESIDENTS *such as bears (below right) and chipmunks (bottom) hibernate to escape the winter cold, while the capercaillie (below) eats conifer buds and shoots in winter. The Bailey Range in Washington, USA (right).*

MOUNTAIN ZONES

As elevation increases, temperatures drop, which creates distinct zones of vegetation. The lowest slopes of a mountain range may be covered by grassland (see p. 156) and/or broadleaf deciduous forest (see p. 160). The montane zone is dominated by forests of conifers. Stunted trees identify the subalpine zone, while the alpine zone begins where trees cannot grow at all. If the mountain is high enough, the peak will be covered year-round by snow and ice, with little if any life. At lower levels on the leeward slope, warmer, drier conditions (see p. 36) create a scrub zone with drought-resistant shrubs. Downwind from the mountain, the climate may be arid, creating desert (see p. 152).

THE MONTANE ZONE

The position of the montane zone varies with latitude. It starts at 9,000 feet (2,700 m) in the Himalayas, 4,000 feet (1,200 m) in the Sierra Nevada of California, 3,000 feet (900 m) in the western Alps of Europe, and sea level in the Chugach Mountains of Alaska.

Long, cold winters and heavy snowfalls create ideal conditions for evergreen conifers, such as pines, firs, and spruces. By growing close together and having dense foliage, conifers create their own microclimate where winter and summer temperatures are less severe and wind speeds are lower. However, the growing season is shorter because of the shade and the persistence of snow under the trees. Plants have difficulty growing in cold soil, because the organic matter in the soil decays and releases nutrients very slowly. Conifer roots have fungi growing on them, called mycorrhiza, that help the roots absorb nutrients and water.

On forested slopes, creeping snow curves the bases of trees into "snow knees". Avalanches may create treeless corridors.

Food is limited in coniferous forests, and mountain animals, such as deer and many birds, migrate to higher elevations as the weather warms and food becomes more abundant. In fall, these animals reverse their migration, heading to lower, warmer elevations. Animals not adapted to cold mountain winters migrate to the lowest slopes, or hibernate in trees or underground burrows, living on stored fat.

The mountain winds, like the dew and rain, sunshine and snow, are measured and bestowed with love on the forests …

The Mountains of California,
JOHN MUIR (1838–1914), Scottish-born American naturalist and writer

AGAINST THE WIND *Many subalpine trees, such as these whitebark pines (left), are flagged by the wind. Most alpine plants grow near the ground, where they are protected from the wind (above). The guanaco (below), a relative of the llama, has a large heart and lungs to compensate for the lower oxygen levels at alpine elevations. Its thick fur reduces heat loss.*

The Subalpine Zone

The subalpine zone is a transitional area between the lush montane forests below and the harsh alpine zone above. It is characterized by scattered, stunted, and misshapen coniferous trees.

Subalpine trees, such as pines, spruces, and hemlocks, often exhibit flagging, a growth form caused by windblown snow damaging the buds, needles, and branches on the windward side of the tree. At the higher end of the subalpine zone, trees grow in prostrate thickets, called krummholz, where the branches protect each other from the severe wind.

Animals in the subalpine zone are a mixture of the animals found in the alpine zone above and the montane below. A mountain-dwelling animal can easily migrate to a more favorable climate. Many montane animals move to the higher elevation subalpine and alpine zones in summer, and retreat down the mountain to more protected areas during winter. The ibex, a goat of the European Alps, lives in the alpine area in summer and the subalpine zone in winter. Year-round residents of the subalpine zone include the yellow-bellied marmot and the alpine chipmunk.

The Alpine Zone

The alpine zone has many of the extreme climatic conditions found in polar climates (see p. 168), such as long, cold winters, strong winds, snow, and ice. And because the atmosphere is thinner at high altitude, the light intensity is much greater in alpine regions.

Snow cover is a variable influence on plants. Too much snow creates a very short growing season and wet, cold soils. Too little snow exposes plants to wind and frost damage in winter, as well as drought in summer.

No trees grow in the alpine zone and most plants are small perennials—plants that live for several years. By staying small and close to the ground, the plants spend less energy growing new tissue in summer and are completely covered by an insulating layer of snow in winter. By being perennial, the plants are able to quickly grow new leaves and flowers

THE PIKA, *a relative of the rabbit, is found in alpine regions of North America and Asia. Its small limbs, ears, and tail limit heat loss. At about 7 inches (17.5 cm) long, it is too small to hibernate.*

BURROWS *Pikas live in rock piles, whistling to each other to warn of predators. During winter, they live on the dried plants they have stored in their burrows.*

in summer from overwintering buds.

Alpine plants tend to have spreading roots and/or long tap roots to absorb scarce water and to remain anchored in strong winds and loose soils. Most alpine plants are hairy to protect them from heat, cold, and the Sun's ultraviolet rays, and to limit the drying effects of the wind. Some plants store all sensitive materials in a bulb or corm underground, safe from the extremes of winter.

Alpine Animals

Very few animals inhabit the alpine zone year-round. Those that do tend to be small because of the scarcity of food. The smallest alpine animals, such as the snow vole of Eurasia, need to eat throughout winter. If they hibernated, they would lose all their fat reserves by metabolizing to stay warm.

Medium-sized animals, such as the alpine marmot, have enough body fat to safely hibernate throughout winter. Larger animals, such as the mountain goat, rely on well-insulated coats.

Birds can easily fly to warmer elevations, but are well suited to alpine climates because of their insulating feathers, lack of external heat-losing appendages, and a circulatory system that delivers twice as much oxygen as a human's. Pheasants in the Himalayas have large, stocky bodies, which minimize heat loss. Other birds, including the white-throated swift of America, are strong fliers and search wide areas for food. Many alpine birds, such as the kea of New Zealand, build nests in holes in the ground where it is warmer.

Insects are small enough to live in protected niches inside leaves, flowers, seeds, and bark. Some alpine insects can tolerate solid freezing, while others, such as the braconid wasp larva, avoid freezing by producing glycerol, a natural antifreeze.

MOUNTAIN MAMMALS *range from the yellow-bellied marmot of subalpine zones (left) to the mountain goat of alpine altitudes (right).*

POLAR CLIMATES

Some species of flora and fauna have adapted well to the frozen landmasses and extreme cold of polar regions.

Centered on the North and South poles are the coldest places on Earth: the Arctic and the Antarctic. These regions are subjected to long periods of darkness and light, and are dominated by snow and ice.

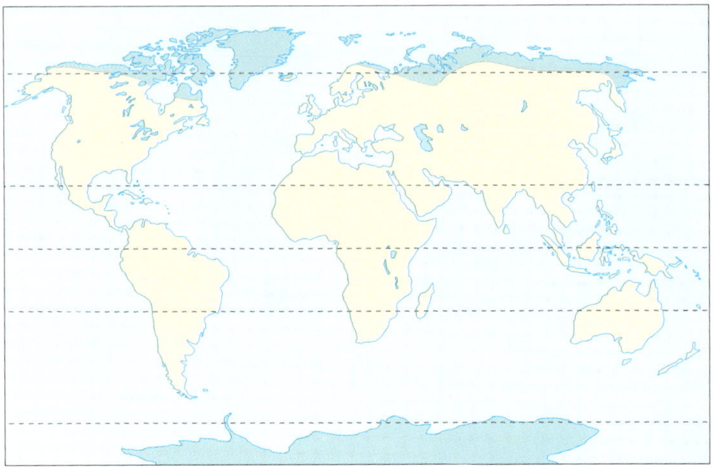

THE ARCTIC CLIMATE

The Arctic includes Greenland and parts of Eurasia and North America, as well as large expanses of ice-covered ocean. The landmasses are dominated by tundra, a treeless plain of low-growing plants adapted to the climate's low temperatures, short growing season, and low precipitation.

HAIR AND FEATHERS *The musk ox (above) has guard hair that, at 24 to 36 inches (60 to 90 cm) long, is the longest hair of any animal. The ptarmigan (top left) is one of the most heavily feathered birds, and even its toes and nostrils have feathers.*

Winter days offer little or no sunshine, and even in summer the Sun stays low in the sky. Temperatures climb above freezing for only two to four months of the year, and even during July, the warmest month, the average daily temperature is no more than 50° F (10° C).

Strong winds carrying ice crystals blow for much of winter, and the thin layer of topsoil is constantly freezing and thawing, making it hard for plants to take root and grow. What little precipitation occurs—about 10 inches (250 mm) a year—is usually in the form of snow or ice. Moisture in the liquid form that animals and plants can use is very limited.

Below the topsoil, the ground is permanently frozen. This permafrost, as it is known, prevents plant roots from penetrating deeply into the soil. It also restricts drainage, creating shallow lakes and bogs.

FALL COLORS *Chemicals that trap heat give many tundra plants colorful leaves in fall, although a few species change color because they are deciduous.*

ARCTIC PLANTS

Because of the short growing season and the severe frosts that occur even in summer, most arctic plants are hardy perennials, such as grasses and sedges (see p. 166). As in the Antarctic, mosses and lichens flourish because they can tolerate freezing temperatures (see p. 170).

THE ARCTIC AVENS, *a widespread tundra plant and the floral emblem of Canada's Northwest Territories, survives extreme cold, strong winds, and a short growing season.*

It was a world of glass,

sparkling and motionless

... Everything was rigid,

locked-up and sealed.

Laurie Lee (b. 1914),
English author

Some plant species grow in a dwarf form in the Arctic. Willows, manzanitas, and birches reach a height of 3 to 60 feet (1 to 18 m) in warm climates, but here grow only a few inches tall.

Arctic Animals

Many arctic animals escape the harsh winter by migrating. More than 120 bird species, particularly waterfowl and shorebirds, spend summer feeding upon the abundant insect and plant life and then depart for warmer climes as winter approaches. Some mammals, such as caribou and reindeer, undertake shorter migrations to find food.

Those animals that are not equipped for long journeys, such as voles and lemmings, take refuge under the snow, where they are sheltered from the worst of the weather and can feed on buried plants.

One resident bird, the ptarmigan, flies directly into soft snow banks to sleep, leaving no tracks for predators to follow. The ptarmigan, as well as other arctic animals such as the snowy owl, the snowshoe hare, and the arctic fox, replaces its brown summer coat with a white winter coat. Changing colors with the season allows these animals to blend in with their environment, and so go unnoticed by predators or prey.

Only the strongest animals can face the winter head on, out in the open. These tend to be large, warm-blooded creatures, such as bears, musk oxen, and wolves, which lose less heat because they have a small surface area in relation to their body size. They are also well insulated by a thick layer of fat and dense fur.

POLAR BEARS *have a counter-current mechanism that limits heat loss. Warm blood from the heart transfers heat to cool blood returning from the skin.*

ANTARCTICA *is covered with an ice sheet (left) that contains 80 to 90 percent of the Earth's fresh water. Crabeater seals (below) live on floating islands of ice, venturing into the water to feed on krill. Lichens (bottom) do not freeze until below −4° F (−20° C) because of a high concentration of proteins and acids.*

Polar Climates of the Antarctic

The Earth's coldest, driest, and windiest climate is found in Antarctica, a continent situated almost entirely within the Antarctic Circle at 66.5 degrees south latitude. The interior has average temperatures ranging from -67 to -76° F (-55 to -60° C). Precipitation, which falls only as snow, averages 2 inches (50 mm) per year. Winds can reach hurricane velocities well in excess of 100 miles per hour (160 kph).

Life can exist on Antarctica only in the milder zones along the coast or on the Antarctic Peninsula. Temperatures will climb a few degrees above freezing during the warmest summer month, while winter temperatures drop to about 0° F (-18° C). Rain averages 16 inches (400 mm) per year.

Antarctica is covered with an ice sheet that averages 6,600 feet (2,100 m) thick. If all this ice ever melted, it would raise the world's sea level by 200 feet (60 m).

Resilient Plants

Only a few plant species can survive the Antarctic climate, and these are found only on the warmer Antarctic Peninsula. The dominant plants are algae, lichens, and mosses—plants adapted to a limited water supply and scant nutrients and soil. These plants remain dormant for most of the year, and their slow growth is best measured in centuries. The only two flowering plant species are the Antarctic hairgrass, a small grass rarely growing above 2 inches (5 cm) tall, and the Antarctic pearlwort, a compact cushion plant with tiny, succulent leaves.

Lichens are composed of an alga and a fungus in a mutually beneficial relationship. The fungus provides nutrients, moisture, and protection, which allow the alga to photosynthesize at a lower temperature than other plants. Lichens don't need soil to grow because they extract all their nutrients from the air and bare rock. As they grow, they release acid that dissolves rock into sandy soil, creating a place where mosses can grow.

Antarctic Fauna

The ocean around Antarctica provides a rich food source for many migratory animals. Birds that visit Antarctica to nest include albatrosses, gulls, petrels, terns, and sheathbills. Like most marine birds (see p. 172), these birds waterproof their feathers with oil from a gland at the base of the tail.

Antarctic waters are home to 15 whale species and 6 seal species. The Antarctic fur seal has two types of hair—

SUB-ZERO *Residents of Antarctic waters range from seaspiders (left) to seals.*

water-repellent guard hairs, and underfur hairs that trap a layer of insulating air.

Cold water contains more dissolved oxygen than warm water, so Antarctic fish have less hemoglobin than other fish. (Hemoglobin is the red pigment in blood that processes oxygen.) The ice fish has no hemoglobin at all, giving it a ghostly appearance.

Among terrestrial animals, the only year-round residents are tiny invertebrates, which feed primarily on algae, fungi, and rotting vegetation. The invertebrate fauna include nematodes, rotifers, tardigrades, midges, and springtails.

Some invertebrates starve and dehydrate themselves during cold weather because any water left in their bodies would encourage deadly ice crystals to form. Others produce antifreeze chemicals that help them survive temperatures as low as -30° F (-35° C).

Springtails are the most common insect in Antarctica. They lay eggs whenever the temperature rises above freezing. In a warm climate, springtails may live for only a few months, but here their metabolism slows down and they can live for two years. This longevity is a common Antarctic adaptation and it allows them another chance to breed if the weather is too cold during the first year.

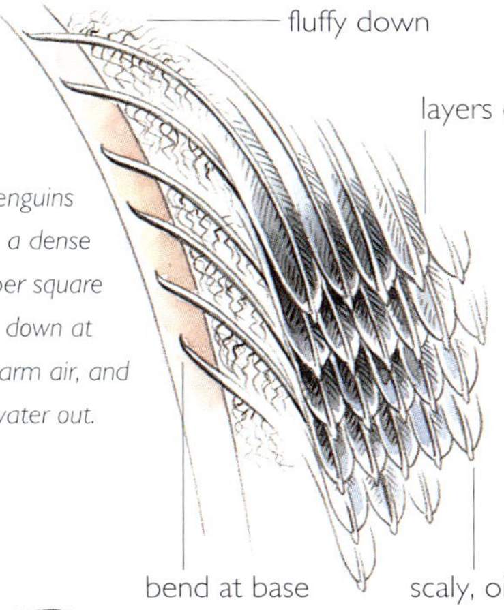

FOR INSULATION, *emperor penguins have a thick layer of blubber and a dense cover of more than 77 feathers per square inch (12 feathers/cm²). The fluffy down at the base of each feather traps warm air, and the scaly, oily tips keep cold seawater out.*

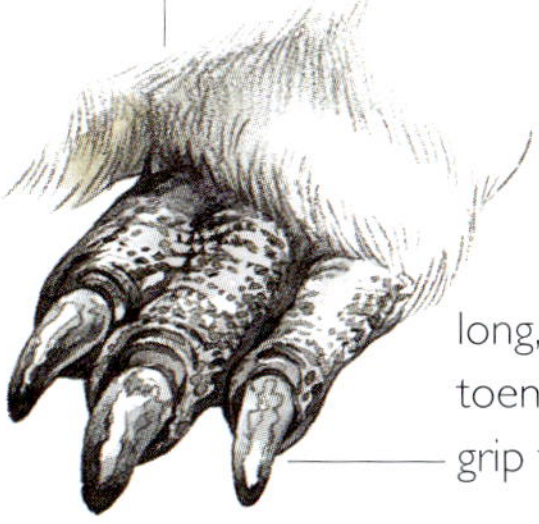

THE ARTERIES *and veins in the feet and flippers are close together, so blood is warmed when it returns from these extremities.*

ANTARCTICA'S EMPEROR PENGUINS *can survive colder weather than any other animal on Earth. They spend weeks in total darkness in temperatures as low as −80° F (−62° C) with winds up to 120 miles per hour (190 kph), while maintaining their internal body temperature at 100° F (38° C). They huddle together for warmth in large colonies, reducing their heat loss by 50 percent. Eggs are laid in the dead of winter, so that the chicks mature in the summer months.*

Coastal Climates

Where land meets ocean, animals and plants must survive the drying effects of strong winds and a salty environment.

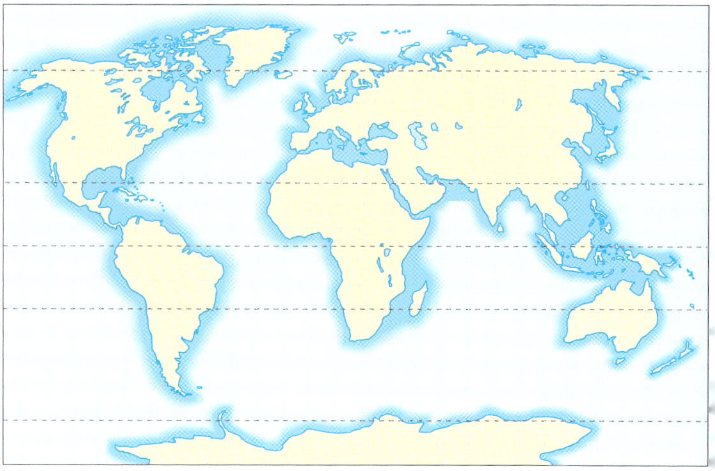

Temperatures tend to be stable along the coast. Oceans respond more slowly to changes in temperature than does land. Sea breezes cool the air on hot days and warm the air on cold days. This does not necessarily create a mild climate on the land. Along the shores, surface temperatures can be very high, and coastal regions are often subjected to waves, fog, and strong winds that carry salt spray.

Sandy Shores

Among the harshest coastal environments are sandy beaches and dunes. Water quickly percolates through the sand creating desert-like conditions. Wind and water carry salt and abrasive sand.

Dune plants have small, waxy leaves to prevent water loss, and large root systems to absorb limited water and anchor the plant in unstable, windblown sand. One of the first plants to grow on new dunes is beachgrass, which has a fast-growing root system and narrow, tough leaves. Other dune plants, such as sea rocket and ice plant, have succulent tissues that store water.

Dune animals must shelter from surface temperatures that can reach 120° F (49° C). Animals such as mice, rabbits, lizards, and tortoises retreat to shallow burrows under shrubs. Winged insects, such as digger wasps, can fly to the cooler air above the sand. Tiger beetles have insulating hairs on their legs, allowing them to walk on hot sand.

Rocky Shores

Rocky shores are exposed to pounding surf, strong onshore winds, intense sunlight, and dry periods at low tide. Just above the high-tide line live blue-green algae and lichens, plants that can tolerate drying out. The periwinkle snail grazes these plants and survives dry periods by retreating to crevices and sealing its shell with a door called an operculum.

Higher up on the cliffs, plants need to be drought-resistant. Some are succulents, such as live-forever. Others have well-adapted leaves: sea

BY THE SEA *Mangroves (above) carry oxygen to their roots via air tubes. The saltmarsh harvest mouse (right) drinks seawater. The ice plant (top) excretes salt via glands that look like ice crystals.*

STRONG WINDS *on the cliffs of Lundy Island, United Kingdom force plants to grow low to the ground.*

THE BROWN PELICAN *of the Americas has adapted to both heat and cold. It can safely fluctuate its body temperature by a few degrees. It loses excess heat by panting and submersing its feet in water. On cold days, it shivers to generate body heat. Its large body helps to conserve heat, and its feathers provide insulation.*

Fog Drip

Sea fog that travels inland may condense onto the leaves of trees, falling to the forest floor as fog drip. This extra moisture can create temperate rain forest, as found in North America's Pacific Northwest.

thrift has thick, narrow leaves, while coastal buckwheat has hairy leaves.

Many seabirds nest on cliffs because they offer protection from predators. The onshore winds also create updrafts that the birds use for easy take-off.

Estuaries

Estuaries form in shallow, protected areas along the coast and are among the world's most productive ecosystems.

Mangrove trees are the predominant vegetation in tropical estuaries. To counter the dehydrating effects of saltwater, mangroves have thick, succulent leaves with a waxy coating, and excrete salt through their leaves. These plants need stable high temperatures and over 75 inches (1,900 mm) of rain per year.

The mangrove canopy protects animals from the sun. The trees block the wind, creating high humidity that allows marine animals such as hermit crabs to venture onto the branches without drying out.

The saltmarsh grasses of temperate estuaries have adaptations similar to those of mangroves, but can also tolerate cold weather.

… every sky has its beauty, and storms which whip the blood do but make it pulse more vigorously.

The Private Papers of Henry Ryecroft,
George Gissing (1857–1903), English novelist and essayist

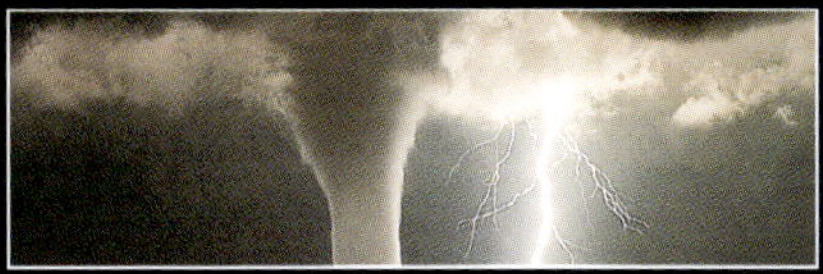

Chapter Eight

The Weather *in* Action

USING THE WEATHER *in* ACTION

This field guide to the most common weather phenomena is the perfect place to start honing your weather-watching skills.

The following pages describe 57 of the most common weather phenomena. Each entry explains where you are likely to encounter the phenomenon, how and why it occurs, and what kinds of weather pattern it is generally associated with.

Most amateur weather-watchers are keen to make use of such information to predict the weather, so, where possible, we have noted if the presence of a certain phenomenon can indicate future weather. It must be emphasized, however, that weather prediction is like assembling a several-hundred piece jigsaw puzzle, and establishing a connection between two or three pieces will provide only a very limited view of the overall picture. Noting, for example, that a certain type of cloud is visible seldom provides enough information by itself to produce a meaningful forecast. However, such an observation may help confirm or place in doubt official forecasts from local weather centers.

It should be remembered that in a field guide such as this one we are attempting to impose an ordered, artificial, classification onto fluid, three-dimensional, and ever-changing phenomena. Many weather patterns will be observed that don't obviously match any of our categories. However, learning to recognize the most common phenomena will help you to interpret and identify more unusual weather patterns.

CATEGORIES OF WEATHER

The Weather in Action is divided into the following categories of phenomena:

Dew, Fog, and Frost all result from condensation or sublimation at or near ground level (see p. 40).

Clouds form when condensation or sublimation takes place above the ground (see

*The **illustrated banding** identifies the category of weather phenomena—see the key above.*

Weather Watch notes:

*✦ **Weather symbol:** where applicable, we have included the relevant meteorological symbol. For a list of symbols, see p. 85.*

*✦ **Distribution:** where, around the world and locally, you are likely to encounter this type of weather.*

*↕ **Height:** the altitude in the atmosphere at which the phenomenon normally occurs. The figures represent typical cases rather than all possible occurrences.*

*◉ **Cause:** the conditions or circumstances that give rise to the phenomenon.*

*➤ **Associated weather:** Other weather likely to occur in conjunction with or as a result of this phenomenon.*

*⚠ **Hazard warning:** dangers associated with this type of weather.*

Dew, Fog, and Frost 178

Clouds 188

Precipitation 218

Storms 236

Optical Effects 254

p. 42). The entries are grouped into low-, middle-, and high-level clouds. Cumulonimbus clouds have been included with the low-level cumuliform clouds because they develop out of such formations.

Precipitation includes the most common types of precipitation, and the effects of too much and too little rain or snow—flood and drought. Since precipitation is defined by the form it is in when it reaches the ground (see p. 46), no height measurements appear in the Weather Watch box in this section.

Storms covers the most extreme and destructive weather phenomena, such as thunderstorms, lightning, tornadoes, and hurricanes.

Optical Effects are really atmospheric phenomena rather than weather. However, these effects not only delight us, but also help us interpret and, on occasion, predict weather patterns.

Fog

level and result in the fog stratus remaining intact for most of the day. This, in turn, will keep temperatures low at ground level.

A thick layer of fog stratus may produce light drizzle, or snow in cold temperatures. However, because the fog tends to get thinner as it rises, any precipitation is generally short-lived.

Extensive areas of fog stratus may obscure large areas of terrain, creating a significant aviation hazard. This can cause problems for the motorist as well, particularly in mountain areas. Valley roads may be clear, but as the motorist ascends he or she may suddenly encounter thick fog, before emerging once again into bright sunshine above the layer of fog stratus.

Fog stratus in the Grand Teton Range, Wyoming, USA, (above), and over vineyards in the Sonoma Valley, California, USA (below).

The main photograph shows a typical view of the phenomenon. Where the image shows a particular variation, this is explained in the caption.

The text provides detailed information on where and when you are likely to encounter the phenomenon, how and why it occurs, and what weather patterns it is normally associated with.

Secondary photographs show further variations, associated weather, and historical images relating to the phenomenon.

Color field sketches supplement the text by illustrating the process that creates the weather phenomenon.

Dew, Fog, and Frost

Dew

Often, after a cold, cloud-free night, we wake to find the ground and other surfaces wet and glistening in the sunlight. The cause of this is a form of condensation called dew. This occurs when the temperature of the ground, or any other surface, drops low enough to cause condensation in the air immediately above it. This results in the formation of water droplets on the surface.

Virtually the same process can give rise to fog (see p. 181), and it is often difficult to predict when dew and fog will occur together. It is possible to have dew without fog, but it is not possible to have fog without dew.

The ideal conditions for dew are a still, clear night; high humidity in the air next to the ground; and low humidity in the air above. The absence of cloud allows the ground to radiate much of the heat it has absorbed during the day and cool sufficiently for condensation to occur. The layer of moist air at ground level ensures that condensation will take place only on, or near, the surface of the ground. With fog, a deeper layer of moist air is required.

In the case of dew, liquid forms because water droplets merge more readily on solid surfaces. In the air, the droplets tend to bounce off each other.

Dew is often associated with cold environments, but it also occurs in hot and humid regions. In desert areas, dew formation is a vital source of water for many plants and animals.

WEATHER WATCH

Worldwide; most common in coastal and tropical areas

On ground, and on surface of grass, leaves, and other objects

Condensation occurring in a thin layer of air immediately above the ground

None

The desert-dwelling Namibian darkling beetle depends on dew for its water supply.

After a cold night, this European skipper butterfly will have to wait for its body temperature to rise again before it can fly.

Radiation Fog

Fog is really cloud that forms near the ground, and, like cloud, fog forms as a result of condensation (see p. 40). Probably the most common form of fog is radiation fog, so named because it is produced by radiational cooling of the ground. This happens at night, when heat absorbed by the Earth's surface during the day is radiated back into space. The highest degree of radiational cooling occurs on clear nights, when there are no clouds to reradiate the heat back to Earth.

Radiational cooling produces condensation in the air layers immediately above the ground. If only a thin layer of moist air is present, dew (see p. 180) will form; if a thicker layer is present, radiation fog (and dew) will form.

Radiation fog varies in depth from only 3 feet (1 m) to about 1,000 feet (300 m). As it is always found at ground level, the most obvious effect of this type of fog is a reduction in visibility, which may drop to as low as 10 feet (3 m) in thick fog. If visibility is between half a mile and 1¼ miles (1 and 2 km), the fog is known as mist. If any smoke is present, it may combine with the fog to produce smog (see p. 266).

WEATHER WATCH

Worldwide

0–1,000' (0–300 m) deep

Cooling of ground causing condensation in air layers near surface

Drizzle or light snow

Restricted visibility

Fog usually disperses soon after sunrise, as the Sun's rays gradually warm the ground. Because clear skies are required for radiation fog to occur, a fine day normally follows. In some cases, however, middle-level cloud may slide over the fog early in the day, inhibiting the clearing of the fog by the Sun.

Dense radiation fogs have caused many aviation and motoring accidents over the years. Even today, despite sophisticated onboard navigation equipment, aircraft landings in thick fog are not normally permitted.

A dense city fog, depicted in Liverpool Docks *by the English painter John Atkinson Grimshaw (1836–93).*

Advection Fog

Advection fog often looks like radiation fog (see p. 181) and is also the result of condensation. However, the condensation is caused not by a reduction in ground temperature, but by moist air drifting into a cold environment (or cold air moving into a moist environment). This means that advection fog can sometimes be distinguished from the normally stationary radiation fog by its horizontal motion. Since radiation fog almost always forms at night, any fog forming during the day is likely to be advection fog.

WEATHER WATCH

Worldwide; most common at sea and in coastal areas

0–1,000' (0–300 m) deep

Moist air moving into cold environment; or cold air moving into moist environment

Drizzle or light snow

Restricted visibility

A satellite image of a bank of sea fog over the North Sea (above). Advection fog fills San Francisco Bay (top).

Sea fogs are always advection fogs, because the oceans don't radiate heat in the same way as land and so never cool sufficiently to produce radiation fog. Fog forms at sea when warm air associated with a warm current drifts over a cold current and condensation takes place. Sometimes such fogs are drawn inland by low pressure, as often occurs on the Pacific coast of North America.

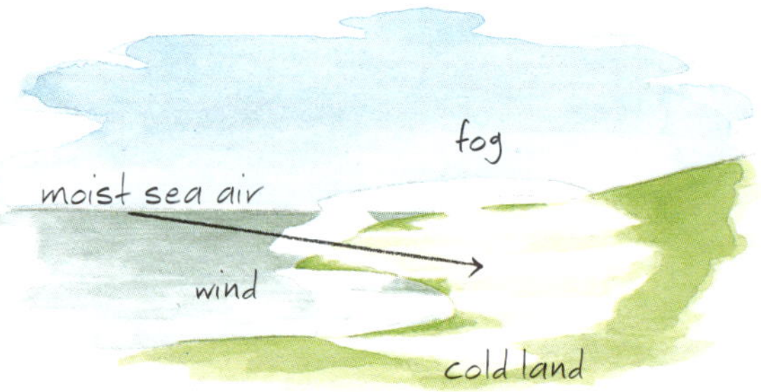

Advection fog may also form when moist maritime air drifts over a cold inland area. This usually happens at night when the land temperature drops as a result of radiational cooling.

Another common form of advection fog is valley fog. In this case, air that has cooled (and thus become denser) during the night drains into a valley from surrounding hillsides. Condensation then takes place, and the valley fills with fog.

Upslope Fog

Upslope fog occurs when moist air is blown up a hillside or mountainside to a level where condensation occurs. The differences between this fog and orographic stratus (see p. 192) are minor. Generally, stratus results from a significant wind, whereas the air currents that produce upslope fog are weak—indeed, someone standing within the fog may be unaware of any air movement. Orographic stratus is more likely to form near the top of a peak, or just above it, whereas upslope fog usually begins farther down the mountain and covers a wider area.

Upslope fog is common in all mountain ranges. Good examples occur in North America during the winter months, when cold air from low-pressure systems drifts slowly westward in the wake of a cold front. When it encounters the eastern slopes of the Rocky Mountains, the air rises and condensation occurs, causing extensive areas of upslope fog, which may run for hundreds of miles in a north–south direction.

A similar process takes place in eastern Australia, when maritime air from the Tasman Sea slides inland and is lifted by the gentle slopes of the Great Dividing Range, creating extensive fog on the eastern side of these mountains.

WEATHER WATCH

Worldwide; most common on hills and mountains near the sea

0–1,000' (0–300 m) above ground level

Gentle lifting of moist air, followed by condensation

Drizzle or light snow

Restricted visibility

As light breezes push moist air up a mountain, it may gradually condense into fog.

Fog Stratus

Normally, a bank of fog forms during the night and begins to disperse as the Sun rises and warms the atmosphere. In certain conditions this can give rise to a bank of fog at a higher level. This phenomenon is known as fog, or low, stratus.

The Sun's rays first heat the ground near the edges of the fog, causing the perimeter to dissipate. Some of the heat also penetrates the deck, warming the ground underneath. The heat from the ground then begins to evaporate the fog at low level. Thus, the fog erodes from the edges toward the center and from the underside up, resulting in a layer of fog some distance off the ground.

To someone standing within the fog, this process is usually signaled by a gradual increase in visibility at ground level. However, if the deck remains intact as it rises, the Sun may stay hidden for some time.

WEATHER WATCH

- *Worldwide; most common in inland areas*
- *0–2,000' (0–600 m) deep*
- *Lifting and erosion of a fog bank by solar heating*
- *Drizzle or light snow*
- *May mask terrain and restrict visibility*

Up and Away

Because fog stratus forms in still conditions, there is seldom any wind at ground level and the stratus normally lifts more or less vertically as it erodes. On some occasions, however, light winds may develop soon after the bank of stratus has formed, and gently blow the fog across the ground as it is dispersing. This will tend to speed up the clearing process.

Generally, fog stratus will have cleared by mid- to late morning, although exceptionally thick layers may take longer to disperse. Sometimes, an increase in middle- or high-level cloud can inhibit the heating process at ground

A deck of fog stratus over a valley in the Swiss Alps.

level and result in the fog stratus remaining intact for most of the day. This, in turn, will keep temperatures low at ground level.

A thick layer of fog stratus may produce light drizzle, or snow in cold temperatures. However, because the fog tends to get thinner as it rises, any precipitation is generally short-lived.

Extensive areas of fog stratus may obscure large areas of terrain, creating a significant aviation hazard. This can cause problems for the motorist as well, particularly in mountain areas. Valley roads may be clear, but as the motorist ascends he or she may suddenly encounter thick fog, before emerging once again into bright sunshine above the layer of fog stratus.

Fog stratus in the Grand Teton Range, Wyoming, USA, (above), and over vineyards in the Sonoma Valley, California, USA (below).

Frost

Frost, like fog, tends to occur on clear nights when the absence of cloud allows heat to rapidly radiate from the ground, resulting in a significant drop in temperature. For frost to form, the temperature must fall to below freezing (32° F [0° C]).

True frost, known as hoar frost, occurs when a thin layer of moist air near the ground cools to below freezing and immediately forms ice crystals, without first condensing as liquid (dew). These crystals will coat any cold surface including stone, grass, leaves, berries, and even spiders' webs. Sometimes, hoar frost is so thick and white that it is mistaken for snow.

The ice crystals that result from hoar frost have exquisite, jewel-like patterns that branch outward from the edges of leaves and grass stems. These intricate structures are easy to see when hoar frost forms on window panes. This normally happens on the windows of an unheated house, when the exterior temperature falls to below freezing. Because moisture levels inside the house are higher than those outside, hoar frost crystals readily form on the inside of the cold window pane, coating the glass with delightful columns, plates, and feathers of frost.

If condensation takes place and dew forms before the air temperature falls below 32° F (0° C), the water, or dew, simply freezes, forming solid droplets rather than delicate ice crystals. These droplets are a form of ice rather than frost, and they occur in the same way as the ice on puddles, ponds, and lakes.

WEATHER WATCH

Worldwide, though only at high altitude in tropical areas

Mainly on the ground, but also on plants, trees, buildings, and other low structures

Water vapor freezing without first forming a liquid

None

Slippery roads; damage to plants, particularly fruit

Jack Frost, shown here in a nineteenth-century British cartoon, is the personification of frost or cold weather in many cultures. The legend probably has its origins in Scandinavian myths.

The crystalline structures that form on plants and other surfaces (below left) as a result of hoar frost can be examined in more detail when they occur on window panes (above).

Frost Damage

When temperatures fall below freezing, the water within the leaves and stems of plants will freeze. This can cause cell damage in the plants and produce a blackening of the leaves. Although this phenomenon is known in some parts of the world as black frost, it is not always accompanied by a frost. Air with a low dew-point (see p. 40) may cool to below 32° F (0° C) without reaching saturation point, which means that no water vapor is released by the air and no real frost formation can occur.

Both hoar frost and black frost are great enemies of citrus-fruit growers, as buds on fruit trees can easily be damaged or destroyed by frost. This can greatly reduce the quantity and quality of the subsequent fruit harvest.

Various methods are therefore employed to minimize or prevent frost formation in orchards. For example, drums of oil may be burned and large, strategically placed fans used to keep the warm air circulating around the trees. Fine water sprays, which have a heating effect because liquid water has a higher temperature than frost, may also be used. Some fruit growers even go to the trouble of hiring a helicopter to fly over the orchard throughout the night. This keeps the air circulating through the trees and provides a small amount of heat from the engine exhausts.

Clouds

Stratus

Stratus clouds form in sheets or layers (*stratus* is the Latin word for layer) and occur when relatively large areas of moist air rise gently in a stable atmosphere to a level where condensation occurs. Normally, the lifting of the air mass is a result of an incoming frontal system or wind encountering a large landmass such as a mountain range.

A slightly different form of stratus may occur when a layer of fog that has developed at ground level starts to rise as it is warmed by the Sun. This formation is known as fog stratus (see p. 185).

WEATHER WATCH

Worldwide; most common near coast and mountains

0–6,500' (0–1,950 m)

Lifting of a large air mass, followed by condensation

Drizzle or slight rain, or snow in sub-zero temperatures

Can be an aviation hazard as it sometimes masks terrain

Stratus is the lowest-altitude cloud formation, with condensation occurring anywhere between ground level and about 6,500 feet (1,950 m). Typically, stratus has a ragged, gray appearance, and varies in thickness from a semi-transparent sheet of a few feet to a deck of around 1,500 feet (450 m). Its horizontal spread is usually far greater and may cover hundreds of square miles.

Normally, there is no significant weather associated with this cloud, although light drizzle or rain, or light snow in sub-zero temperatures, may fall if the deck is sufficiently thick. Stratus that gives rise to precipitation is often known as nimbostratus.

When it forms close to the ground, stratus can mask the surrounding terrain, particularly in mountainous areas, and this has been the cause of many aviation accidents. Fortunately, radar equipment fitted to most modern aircraft is reducing the dangers associated with this cloud.

This illustration from Charles Blunt's Beauty of the Heavens *(1849) shows stratus forming close to the ground.*

Stratocumulus

One of the most common clouds worldwide, stratocumulus is a good indicator of moisture in the lower levels of the atmosphere. It usually occurs between 2,000 and 6,500 feet (600 and 1,950 m).

Stratocumulus usually has a ragged appearance along its upper surface, but can have a well-defined and flattish base. It tends to form in comparatively shallow layers, sometimes several hundred miles wide. The color of the cloud may vary from white to dark gray, depending on the light conditions and the thickness of the deck. Its somewhat lumpy appearance, indicative of convection within the cloud, is what distinguishes stratocumulus from stratus.

Two processes may give rise to stratocumulus, either separately or in combination. In the first, a large, moist air mass is lifted by a frontal system or a landmass to a level where condensation occurs; slight instability at cloud level then creates the cloud's cumuliform shape.

The second process involves pockets of warm air rising from the ground as a result of weak convection, giving rise to condensation at the same level over a wide area. In this case, the clouds may subsequently develop into cumulus humilis (see p. 194), and some of these may even develop into cumulonimbus clouds.

Normally, if stratocumulus has not developed vertically by mid-afternoon, the time of maximum ground temperature, it will tend to dissipate, resulting in a clear evening sky. If it is sufficiently thick, stratocumulus may produce light drizzle, or snow in sub-zero temperatures, but this is not common.

WEATHER WATCH

Worldwide

2,000–6,500' (600–1,950 m)

Lifting of a large air mass, followed by condensation combined with relatively weak instability at cloud level

Normally none, but may produce light precipitation if cloud is sufficiently thick

Landscape with a Ploughman, *by Théodore Rousseau (1812–67), depicts a typical stratocumulus formation.*

Orographic Stratus

Orographic clouds are formed when moist air, carried on a prevailing wind, is lifted by an elevated landform, such as a mountain range, to a level where condensation takes place. Among the most common orographic clouds is orographic stratus. This low-level formation occurs most frequently in areas, such as coastal regions, where the air flow is heavily moisture laden. Generally, the landform must be at least 500 feet (150 m) high to generate cloud, and higher in areas of clean, dry air, such as deserts.

WEATHER WATCH

Over hills and mountain ranges worldwide

0–1000' (0–300 m) above ground level

Lifting of moist air mass by landform

May produce fog, light drizzle, or light snow showers

Can be an aviation hazard as it masks terrain

Unlike normal stratus (see p. 190), which is carried about by the external wind field, orographic stratus tends to remain stationary. The wind flows through the area of condensation, constantly regenerating cloud as the air rises and dissipating it as the air descends on the other side of the landform.

The extent of this type of cloud depends on the humidity of the surrounding air mass. If the air contains a high level of moisture, cloud can begin to form well down the windward slope of the landform, wrap around the peak, and extend some distance down the other side. A good example of this type of formation is the "tablecloth" that often drapes the top of Table Mountain near Cape Town in South Africa.

The extent of the cloud is also determined by the steepness and elevation of the landform, the strength of the wind, and the direction of the wind relative to the landform. Strong wind blowing at right angles to a steep mountain will create greater uplift and generate more cloud.

Since low-level moisture and high land are essential ingredients in the formation of orographic stratus, areas that have high humidity

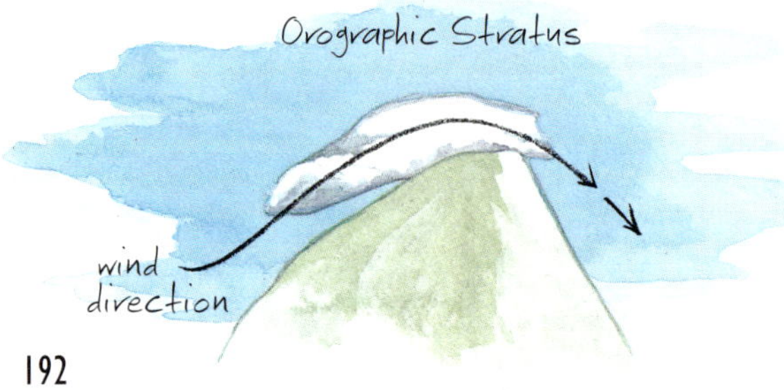

and steep terrain—tropical islands such as Hawaii, for example—are particularly conducive to the formation of these clouds.

Sometimes, there is insufficient moisture in the lower layers of the atmosphere to allow condensation at ground level, but with increasing altitude and lower temperatures, condensation may be possible. In this case, lifting of the air mass by a landform can produce middle-level orographic formations, known as lenticular clouds (see p. 210).

Banner Clouds

One of the most spectacular forms of orographic stratus occurs in a variation known as banner cloud (or cap cloud), which can form on the peaks and immediately to the lee side of large mountain ranges. In this case, the cloud-producing mechanism is a little different.

When wind blows over a tall, steep mountain, a certain amount of air stalls and piles up on the windward side, resulting in an area of increased air pressure. This, in turn, results in a drop in air pressure immediately to the lee side of the mountain. Since low pressure enhances con-

densation, a cloud will readily form in this position if sufficient moisture is present.

Two celebrated examples of banner cloud stratus are the mighty plumes that sometimes trail the peak of Mount Everest, in the Himalayas, and the Matterhorn, in the Swiss Alps. However, this process may also be seen in a weaker form near lesser mountain peaks.

Orographic stratus over Table Mountain, Cape Town, South Africa (above). Banner cloud in Nepal (right).

Cumulus Humilis

Cumulus clouds generally form as a result of localized pockets of warm air rising. The water vapor in the air condenses into well-defined, lumpy parcels of cloud at low level. The shape of these clouds gives rise to the name *cumulus*, which means heap in Latin.

Cumulus humilis is the smallest form of cumulus cloud and results from relatively weak convection (*humilis* means humble in Latin). This produces clouds that generally have flat bases and small, rounded tops. Technically, a cumulus cloud is considered to be a humilis formation if it is wider than it is tall, as estimated by an observer on the ground.

The level of the base of this small cloud is determined by the humidity of the surrounding air mass. In areas of high humidity, such as coastal and tropical regions, the base may start at around 2,000 feet (600 m), but it may begin at much higher levels in dry areas.

WEATHER WATCH

Worldwide, except Antarctica

2,000–3,500' (600–1,050 m)

Weak convection

None

Often, cumulus humilis represents an early stage in the development of a cumulus cloud from stratocumulus (see p. 191) to cumulus mediocris and then cumulus congestus (see pp. 195–6). If conditions are right, some humilis clouds may eventually develop into cumulonimbus incus (see p. 200).

Because of its shallow depth, this cloud formation does not produce any significant weather. It can, however, give rise to turbulence during aircraft penetration of the cloud, but this effect is normally slight and short lived.

Cumulus humilis is widely distributed, occurring over all landmasses and oceans when conditions are right, except Antarctica, where the cold surface temperatures inhibit convection.

Cumulis humilis is also known as fair-weather cumulus.

Cumulus Mediocris

Cumulus mediocris is generated by slightly stronger convection than that which gives rise to cumulus humilis. This produces a cumulus cloud as tall as it is wide, as estimated by an observer on the ground.

The base of this medium-sized cloud (*mediocris* means moderate in Latin) can begin to form from 2,000 feet (600 m) upward, depending on the surrounding humidity. Cumulus mediocris is normally white or light gray, and has a comparatively flat base.

This cloud is often a transitional stage between the lesser humilis and the more developed congestus phase (see p. 196). Mediocris clouds are more common in the late morning or early afternoon, after the ground has warmed enough to generate convection. They are not large enough to produce precipitation. During aircraft penetration of the cloud, they can produce slight and short-lived turbulence.

If mediocris clouds occur at levels where strong winds are blowing, the winds may shred the clouds into horizontal fragments, which then speed across the sky. This variation is known as cumulus mediocris fractus.

Mediocris occurs over most oceans and landmasses when conditions are right, with the exception of Antarctica, where the cold surface temperatures generally inhibit convection.

WEATHER WATCH

Worldwide, except Antarctica

2,000–4,000' (600–1,200 m)

Weak to moderate convection

None

Wheat Fields, by the Danish artist Peter Hansen (1868–1938), shows cumulus mediocris covering much of the visible sky.

Cumulus Congestus

Cumulus congestus represents the next stage in the vertical development of a cumulus cloud after cumulus mediocris (see p. 195). Powered by strong updrafts, this cloud may grow to an altitude of 15,000 to 20,000 feet (4,500 to 6,000 m). Congestus clouds are taller than they are wide, and have a flat base and a sharp outline.

WEATHER WATCH

Worldwide, except Antarctica

2,000–20,000' (600–6,000 m)

Convection, enhanced by atmospheric instability

Can produce moderate to heavy showers

Moderate turbulence at cloud level

Congestus seldom forms as a result of convection alone. Normally, atmospheric instability is also required. This occurs when the temperature of the surrounding air mass drops more rapidly with height than is normal, often as a result of a cold air current sliding over the cloud (see p. 42).

Congestus may grow to the cumulonimbus stage (see p. 200) if convection is strong enough or the surrounding atmosphere becomes yet more unstable. The time of day may be a deciding factor, as convection over land weakens late in the afternoon when ground temperatures begin to fall. If, by this time, the tallest cloud in the sky is congestus, then it is unlikely to progress to cumulonimbus.

Congestus is capable of generating heavy and prolonged showers of rain or snow. In fact, during winter in North America, congestus clouds forming downwind of the Great Lakes often produce significant snowfalls.

The vigorous convection that produces congestus creates significant levels of turbulence within the cloud. However, while this may result in a bumpy ride for aircraft passengers, it does not pose a serious threat to their safety.

A rain shower falls from a small congestus cloud in this view of Kraków in Poland, painted by J. Silbermann in 1888.

Pyrocumulus

This variation of cumulus derives its name from the fact that fire (*pyro* in Latin) creates both the lifting mechanism and the water vapor that combine to form this cloud.

An extensive wildfire produces vigorous rising air currents and a large quantity of water vapor that is released by the air and vegetation during combustion. The rising air lifts the water vapor to a level where it condenses and forms cumulus clouds that ride above the fire. The bases of these clouds are often difficult to discern, as they are usually hidden by the smoke from the wildfire, but the cloud tops are normally situated well above the smokescreen.

Pyrocumulus clouds vary widely in vertical extent, from humilis to congestus size. In some cases, the cloud can produce rain showers that limit or even extinguish the blaze below. However, particularly in subtropical regions where condensation results from an abundance of moisture in the surrounding air mass, the clouds may continue to grow until they reach the cumulonimbus stage. In this case, lightning strikes from the cumulonimbus may trigger further fire outbreaks.

Pyrocumulus clouds may be seen wherever wildfires occur. They are inevitably more common in highly fire-prone areas such as California, the French Riviera, and southeastern Australia.

WEATHER WATCH

Any area where wildfires occur

2,000–30,000' (600–9,000 m)

Convection created by fire

Can generate showers and storms

Clouds may develop into cumulonimbus and trigger further fires through lightning strikes

Normally, only the top of a pyrocumulus cloud will be visible above the smoke of the fire.

Cumulonimbus Calvus

Cumulonimbus calvus represents a transitional stage between cumulus congestus and a fully fledged cumulonimbus incus (see p. 200).

Calvus occurs when convection and atmospheric instability combine to push the cloud tops beyond the congestus stage to heights of up to 30,000 feet (9,000 m). At this level of the troposphere, temperatures are normally well below freezing, and any condensation that takes place will produce ice crystals rather than water droplets. This gives the top of the cloud a brilliant, white appearance. However, the cloud will not yet have developed the anvil-like profile characteristic of cumulonimbus incus (see p. 200).

Calvus clouds always produce some form of precipitation, with rainshowers occurring in temperate zones and snowfalls in colder areas. Under certain conditions, these falls can be moderate to heavy. In dry areas, showers may fall from the cloud base but evaporate before reaching the ground, a phenomenon known as virga (see p. 220).

The powerful convective updrafts associated with calvus clouds can produce significant turbulence. However, the precipitation associated with the cloud can usually be located by onboard aircraft radar and evasive action taken.

WEATHER WATCH

Worldwide, except Antarctica

3,000–30,000' (900–9,000 m)

Powerful convection, enhanced by atmospheric instability

Moderate to heavy showers, strong winds

Significant turbulence at cloud level

The mushrooming top of the calvus formation is a sign of the vigorous updrafts that may eventually force the cloud up into the highest levels of the troposphere.

Cumulonimbus with Pileus

Once a cloud has reached the calvus stage, and if convection is still occurring and is enhanced by instability in the surrounding air mass, the cloud will continue to grow vertically. When the air is ascending quite rapidly—speeds of 20 to 30 miles per hour (32 to 48 kph) directly upward are possible—a rather curious phenomenon may take place.

The strong updraft associated with the calvus cloud picks up a slab of air and thrusts it upward. This causes the water vapor in the slab to condense, and a smooth, elongated, cap-like formation, known as a pileus cloud (*pileus* is the Latin word for felt cap) appears above the rising mass of the calvus. As the calvus cloud continues to rise, it gradually catches up with the pileus cloud. When the two clouds meet, some of the pileus cloud topples off the sides of the rising calvus. Eventually, the ascending cloud overtakes the pileus, which remains draped over the peak of the calvus until the two clouds merge completely.

Pileus clouds can help weather-watchers anticipate thunderstorms, as the clouds that generate pileus formations are those most likely to develop into full-blown cumulonimbus clouds.

WEATHER WATCH

Worldwide, except Antarctica

20,000–30,000' (6,000–9,000 m)

Strong convection lifting and eventually absorbing a layer of air

None from pileus, but calvus cloud will produce precipitation

Significant turbulence at cloud level

An aerial view of a pileus cloud. If the convection in a cloud is strong enough to produce pileus, it is likely to lead to a thunderstorm.

Cumulonimbus Incus

A mature cumulonimbus incus is definitely the "King of Clouds", a mighty mountain of moisture often considerably taller than Mount Everest, and sometimes reaching 60,000 feet (18,000 m) in tropical and subtropical areas.

In its full magnificence, it is crowned with a huge, wedge-shaped mass of high cloud resembling a blacksmith's anvil (*incus* is the Latin word for anvil).

WEATHER WATCH

Worldwide, except Antarctica; common in the tropics

2,000–35,000' (600–10,500 m)

Powerful convection assisted by atmospheric instability

Heavy rain or hail, strong winds

Severe turbulence in cloud; strong winds, lightning, hail, and even tornadoes at ground level

This part of the cloud, often referred to as a hammerhead or thunderhead, is a clear sign of a fully fledged thunderstorm.

Cumulonimbus incus can begin early in the morning as cumulus humilis, then proceed through the stages of mediocris and congestus (see pp. 194–6). For the cloud to continue developing at this point, the convective process must combine with atmospheric instability to produce a powerful updraft.

As long as the air in the vicinity of the updraft remains unstable, the cloud continues to rise and expand. Eventually the cumulonimbus cloud reaches the top of the troposphere, where the air temperature levels off and begins to increase with altitude (see p. 24). This change in temperature has the effect of placing a lid on the updraft, and

The anvil-like top of a cumulonimbus is a reliable indicator of a mature thunderstorm.

A magnificent example of a cumulonimbus incus, clearly showing its bright cirrus crown.

the cloud can rise no further. However, the momentum of the air below continues to push upward, and spreads the cloud out in a radial fashion at the tropopause, forming the characteristic anvil shape. The position of this formation therefore indicates the height of the troposphere in the area.

Because the anvil is situated well above the level where the air temperature drops below freezing, this part of the cloud is composed of ice crystals, which forms a crown of cirrus above the main cloud mass. These ice crystals may be blown about by strong, high-level winds, producing a streaky appearance.

Warning Signs

In rare cases, the updraft associated with the cloud is so powerful that it punches through the tropopause and carries a parcel of cloud into the lower levels of the stratosphere, before losing momentum and falling back. This produces an upward bulge on the otherwise flat upper surface of the anvil—a good indicator of a particularly severe storm, which may produce hail, strong wind gusts, and even tornadoes.

A cumulonimbus incus formation must always be regarded as a significant aviation hazard because of the powerful air currents involved in its formation and the potentially damaging effect of the large hailstones it may produce. Fortunately, because of the precipitation generated, the cloud is easily located by onboard and ground-based radar, and elaborate procedures are in place to steer aircraft around any such activity.

Cumulonimbus with Mammatus

Mammatus is one of the most spectacular and distinctive of all cloud formations, making it a favorite with weather-watchers and photographers alike. It consists of pendulous globules of cloud (*mamma* is the Latin word for breast) that hang from the underside of the anvil of a thundercloud (see p. 200). Mammatus is always associated with mature cumulonimbus clouds and is therefore an indicator of severe weather conditions.

WEATHER WATCH

Worldwide, except Antarctica; common in the tropics

15,000–25,000' (4,500–7,500 m)

Strong convection followed by reverse-direction convection

Heavy rain, wild squalls, hail

Severe turbulence in cloud; strong winds, hail, lightning, and risk of tornadoes

Mammatus clouds warn of a particularly intense thunderstorm.

The formation occurs as a result of a process that can be described as reverse-direction convection. During a thunderstorm, warm, moist updrafts rise to the top of the troposphere (see p. 24). Here the temperature levels off and the air stabilizes. This causes the rising cloud to expand horizontally over areas of cooler, cloud-free air. The temperature difference between the two air masses creates instability under the anvil, which causes pockets of warm, moist air in the cloud to convect downward. This reverse-direction convection is enhanced by the effects of gravity and by precipitation from the cloud. The process produces near-symmetrical protuberances on the underside of the anvil, known as the mammatus, which may cover large areas.

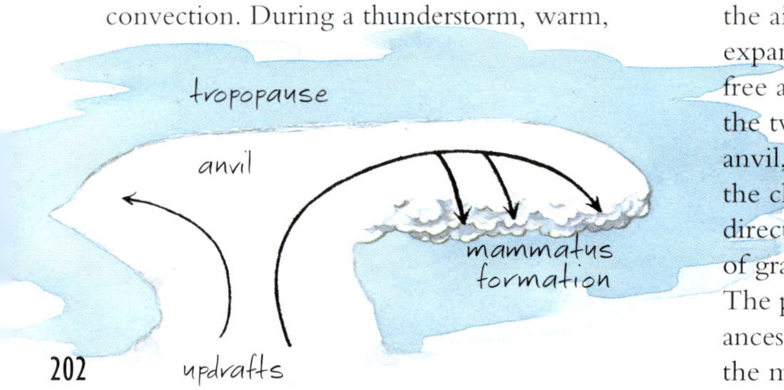

Mammatus forms on the underside of the anvil of a cumulonimbus. The center of the storm may be some distance away.

Tornado Alert

As the anvil of a mature cumulonimbus cloud can spread out over hundreds of square miles, the center of the storm may be some distance away from the mammatus formation. However, mammatus normally occurs soon after the cumulonimbus cloud has reached maximum growth and intensity and is usually a sign of a particularly vigorous thunderstorm. In the United States, this formation is regarded as a clear warning sign of possible tornado development.

Airline pilots will normally take action to avoid any cumulonimbus clouds, but particularly those bearing mammatus formations, as these indicate especially severe turbulence within the cumulonimbus.

Mammatus may be observed wherever cumulonimbus clouds occur, but it is particularly common in areas where thunderstorms are severe, such as tropical and subtropical areas. Because the formation is associated with mature cumulonimbus clouds, it is most likely to be seen from mid-afternoon to early evening, when ground heating and associated convective activity have reached a maximum.

Altostratus

This cloud is found in the middle levels of the atmosphere and is always a sign of the presence of significant amounts of moisture in those layers. It is typically featureless, ranging from a thin, white veil of cloud through which the Sun is plainly visible, to a dense, gray mantle that may block out the Sun completely.

WEATHER WATCH

Worldwide; common in middle latitudes

6,500–16,500' (2,000–5,000 m)

Lifting of a large air mass, followed by condensation

Extensive areas of rain and snow

Ice accretion on aircraft

Altostratus is the result of the lifting and condensation of a large air mass, usually by an incoming frontal system. This can result in an extensive deck of cloud, which may extend over thousands of square miles. If sufficiently thick, altostratus can produce rain or snow over a wide area.

When stratus cloud covers the entire sky, it can be difficult to determine whether it is a low- or middle-level formation. As a rough guide, if you can discern a texture in the cloud deck, it is more likely to be low-level stratus; if it appears smooth and structureless, it is more likely to be an altostratus formation.

For pilots, a thick deck of altostratus can be a cause for concern if temperatures within the cloud are below freezing, because ice may build up on parts of the aircraft as it passes through the cloud, altering the plane's aerodynamics. Fortunately, most aircraft are equipped with de-icing devices that eliminate this problem.

Thick bands of altostratus partially cover the sky in Benjamin Leader's In the Evening It Shall Be Light *(1882).*

Altostratus Undulatus

Altostratus undulatus usually occurs in a thin layer of altostratus, and its distinctive, undulating appearance is due to wave motion in the air mass. This motion is normally a result of wind shear, which occurs when one layer of air slides over another layer moving at a different speed or in a different direction (or both). This creates vertical eddies, or waves, of air between the layers, and if sufficient moisture is present, cloud will form where the wave rises and dissipate where it falls.

Depending on the moisture content of the air mass and the degree of wind shear, undulatus may occur as fairly continuous waves across the sky with thin cloud connections at the base, or it may break up into unconnected wave peaks.

Undulatus produces no significant weather, but, because the cloud is produced by wind shear, it is regarded as a sign of local turbulence. In most cases, however, this turbulence would be only slight and would not be a concern for aircraft pilots.

WEATHER WATCH

Worldwide

6,500–16,500' (2,000–5,000 m)

Lifting of a large air mass, followed by condensation combined with wind shear at cloud level

None

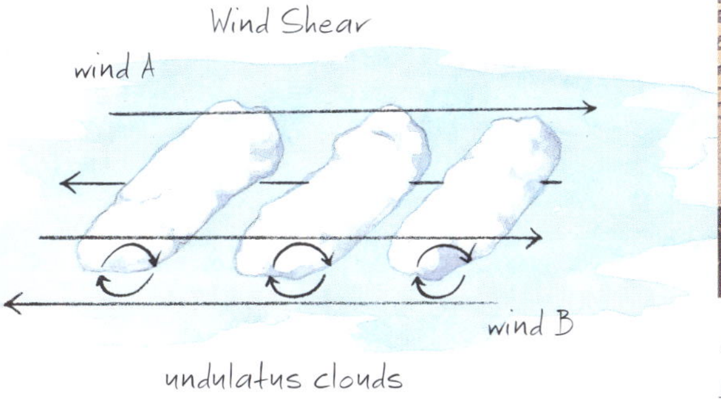

Altostratus undulatus over Mount Fuji, as depicted by the Japanese artist Katsushika Hokusai (1760–1849).

Altocumulus

While altostratus is often flat and featureless, altocumulus usually creates interesting and varied skies. In some cases, many thousands of small altocumulus clouds will be strung together across the sky in spectacular formations.

As with altostratus, altocumulus normally occurs when a large air mass is lifted to middle levels by a landmass or an incoming frontal system, and condensation occurs over a wide area. The principal difference between the two formations is that altocumulus is affected by instability in the surrounding atmosphere. This gives rise to its distinctive cumuliform texture.

In isolation, altocumulus does not have great significance for the weather-watcher, although it can produce light precipitation if the deck is sufficiently thick. However, if the extent of an alto-cumulus formation appears to be increasing during the course of a day, this may be a sign of an approaching frontal system.

Often altocumulus and altostratus appear together in a mixed sky. Satellite photography reveals that mixed altocumulus–altostratus formations may extend over thousands of square miles, particularly when associated with a frontal system.

If altocumulus combines with a thick deck of altostratus at a level where the temperature is below freezing, significant airframe icing may affect the aerodynamics of aircraft flying through the cloud. Otherwise, slight to moderate turbulence will be the only concern for the pilot.

WEATHER WATCH

Worldwide

6,500–16,500'

(2,000–5,000 m)

Lifting of a large air mass, followed by condensation combined with instability

Light rain, if cloud is thick; may indicate an approaching front

Ice accretion on aircraft

Altocumulus formations are often more distinct and dramatic at sunrise and sunset.

Altocumulus Undulatus

Altocumulus undulatus occurs when a layer of altocumulus cloud is affected by wind shear. The mechanism is the same as that which gives rise to altostratus undulatus (see p. 205). The altocumulus variation consists of parallel bands of cumulus clouds. These may form in patches or extend over a wide area of the visible sky. When the bands form close together, they often resemble ripples on the surface of a pond.

Altocumulus undulatus is distinguished from altostratus undulatus by its discernible cumuliform texture. As with all cumulus clouds, this is a result of a certain amount of instability at cloud level, which gives rise to further uplift at various points within the cloud.

These clouds always indicate the presence of significant amounts of moisture at middle levels, and, if on the increase, may signal the approach of a frontal system. If the cloud deck is sufficiently thick, this formation can produce rain, or snow in sub-zero temperatures. Often, altocumulus undulatus occurs together with altostratus formations in a mixed sky. In such cases, it may be difficult to discern which cloud formation is producing the precipitation.

All undulatus clouds are regarded by those in the aviation industry as a sign of turbulence. However, altocumulus undulatus seldom gives rise to anything more than slight to moderate turbulence and is therefore not regarded as a danger by pilots.

WEATHER WATCH

Worldwide

6,500–16,500' (2,000–5,000 m)

Lifting of a large air mass, followed by condensation combined with instability and wind shear at cloud level

Light rain, if cloud is thick; may indicate an approaching front

Altocumulus undulatus is normally the result of wind shear at middle levels of the troposphere.

Altocumulus Mackerel Sky

The mackerel sky variation of altocumulus is named for its resemblance to the scales of a fish. It is almost certain that the name originated among early mariners, and it may have considerable antiquity.

As with altocumulus formations in general, mackerel sky is produced by the lifting of a large, moist air mass, usually by an approaching cold front, combined with instability at cloud level.

WEATHER WATCH

Worldwide

6,500–16,500' (2,000–5,000 m)

Lifting of a large air mass, followed by condensation combined with instability and wind shear

May indicate an approaching frontal system

The exact causes of this pattern have not been firmly established, but it is likely that a form of wind shear, similar to that which produces undulatus formations (see p. 205), is the cause. In this variation, the wind shear gives rise to a more intricate pattern of small waves, which produces the much finer texture of mackerel sky.

Because it is often created by an approaching frontal system, mackerel sky has long been associated in folklore with deteriorating weather conditions. More often than not, this formation, like other middle-level clouds, is a good indicator of changing weather, although, as in all such cases, the front may pass some distance away from the observer, resulting in little change in local conditions.

The silvery scales of mackerel sky.

Altocumulus Castellanus

This cloud is named for the turret-like protruberances that grow from its main deck, resembling the battlements of a medieval castle. While not a spectacular formation, it is significant in that it indicates instability in the middle layers of the atmosphere and may therefore point to thunderstorm development later in the day.

Castellanus generally occurs when a layer of colder air slides across an area of altocumulus cloud. This creates instability, and localized bubbles of air start to rise from the cloud deck. Condensation within these pockets of air creates the castellanus effect.

WEATHER WATCH

M

Worldwide

6,500–16,500' (2,000–5,000 m)

Lifting of a large air mass, followed by condensation combined with instability

May point to thunderstorm development later in the day

Any subsequent convection from the ground will be enhanced by this middle-level instability. Cumulus clouds forming in these areas are therefore more likely to develop into cumulonimbus clouds.

For this reason, meteorologists are always alert to reports of castellanus. If several ground observers have noted castellanus development by the middle of the day, then there is a greater likelihood of thunderstorm activity later in the afternoon, following further ground heating.

Since the formation of altocumulus castellanus involves vertical air currents, pilots can expect to experience slight to moderate turbulence as they pass through these clouds, but this will pose no danger to aircraft safety.

Convective currents create turret-like clouds.

Altocumulus Lenticularis

WEATHER WATCH

Common over mountain ranges worldwide

6,500–16,500' (2,000–5,000 m)

Air mass forced to rise to condensation level by landmass

Light rain or snow; high winds

Moderate turbulence at cloud level

Altocumulus lenticularis clouds are named for their smooth, round, lens-like shape (lenticular means lens- or lentil-like). These middle-level clouds can form spectacular patterns that delight weather-watchers and photographers and have almost certainly been responsible for a number of UFO sightings over the years.

When wind blows across a mountain range, it tends to form air waves on the lee side of the mountains. This process, known as the mountain wave effect, is usually invisible, but when moisture is present at the top of these waves, lenticular clouds form where the wind rises and dissipate where it falls.

Because mountain ranges are nearly always of irregular shape and wind may move at different speeds at different levels, the waves produced in this manner often have varying distances between their crests (referred to as the wavelength) and the resulting clouds form an irregular pattern.

However, if the mountain range has a fairly regular shape and the wind is blowing at a steady speed at approximately right angles to the mountains, the wave crests, and any resulting clouds, will form a regular pattern. Furthermore, if alternate layers of moist and dry air are present above the mountains, the clouds may pile up on top of each other like stacks of plates. It is these distinctive stack formations that have, on occasion, been mistaken for UFOs.

If the wind generating the waves has a fairly constant speed, the cloud pattern will be stable

Lenticular clouds remain stationary as the wind passes through the cloud mass, constantly generating and dissipating the condensation.

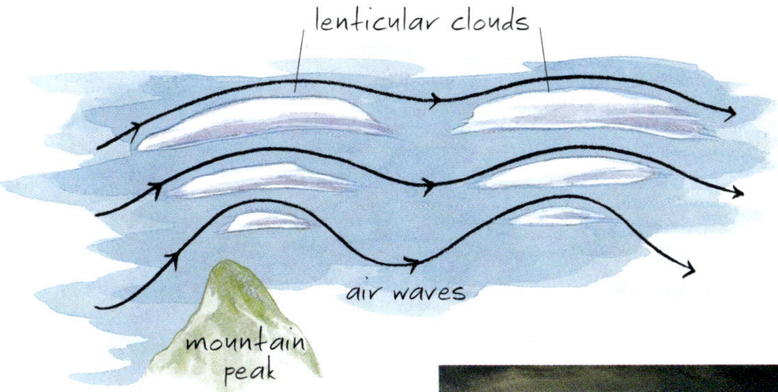

and long lasting, remaining virtually stationary in the sky for extended periods.

Generally, no significant weather is produced by altocumulus lenticularis, but, occasionally, if there is sufficient moisture in the surrounding atmosphere, these clouds can become thick enough to generate light rain, or snow showers in sub-zero temperatures. Because these formations are associated with high-speed winds in the middle layers, they may be precursors of windy conditions at ground level.

A dramatic lenticular cloud over South Orkney Island in Antarctica.

Surfing the Waves

The roller-coaster motion of the air, made visible by the formation of these clouds, can produce significant levels of turbulence, and commercial airliners will try to avoid such areas. High-level glider pilots, however, sometimes do the opposite: they look for lenticular clouds as signs of a source of uplift for the aircraft. The gliders can surf along the mountain waves, maintaining altitude by remaining on the rising side of the wave crest.

Lenticular clouds occur in most parts of the world, and may form over quite small mountains. Good examples occur when moist winds blow in off the Pacific Ocean and encounter the Sierra Nevada in California. However, the most dramatic clouds are generated by the largest ranges, such as the Himalayas, Andes, and Rocky Mountains.

Cirrus

The Latin word *cirrus*, meaning wisp of hair, is the name given to high-level clouds that stretch across the sky in delicate strands. These formations indicate the presence of moisture at high levels of the atmosphere. At these levels, the temperature is normally below freezing, and any air mass that cools to saturation will produce ice rather than water droplets. Cirrus clouds therefore consist of many millions of ice crystals, and these are blown about by upper-level winds, producing characteristic white streaks.

Among the most distinctive cirrus shapes are the irregular twists and tangles of cirrus intortus and the hook shapes of cirrus uncinus (see p. 213). Less common, but equally dramatic, is cirrus radiatus, which forms in long, parallel lines that seem to radiate from a point on the horizon.

Cirrus clouds may form in isolated patches or cover a wide area of the sky, depending on the distribution of moisture. Isolated patches rarely have any great significance, but an extensive deck, increasing from one direction, may indicate an approaching front.

WEATHER WATCH

Worldwide

Above 16,500' (5,000 m)

Saturation of air mass at upper levels

If extensive, may indicate an approaching frontal system; may also be a sign of a decayed thunderstorm

Cirrus may be the result of local thunderstorm activity. The anvil that often forms above a cumulonimbus cloud (see p. 200) is actually a cirrus cloud. It is generated as the thunderstorm pumps moisture up to the very top of the troposphere, where it freezes into ice crystals. After the storm has completed its life cycle, high-level winds may disperse the anvil across the sky, producing extensive cirrus formations far downwind.

Cirrus intortus over Sydney, Australia (above). Cirrus radiatus (right) appears to emerge from a point on the horizon.

Cirrus Uncinus

This often spectacular variation of cirrus cloud is also known as hooked cirrus (*uncinus* is the Latin for hook), or cirrus mares' tails, a reference to the cloud's resemblance to a horse's tail.

Cirrus uncinus forms in much the same way as other cirrus formations. However, its distinctive pattern of filaments is the result of a high-speed wind below the level at which the ice crystals form. As the crystals descend under the influence of gravity, this wind rapidly smears them across the sky, forming the distinctive, elongated, hooked shapes.

Like other cirrus clouds, uncinus is a result of high-level moisture, and is therefore often associated with the approach of a frontal system. Since it is also evidence of a high-speed, high-level wind, it may indicate the presence of a jet stream (see p. 31).

WEATHER WATCH

Worldwide

Above 16,500' (5,000 m)

Saturation of air mass at upper levels, combined with strong wind immediately below cloud level

May indicate an approaching frontal system

Normally, cirrus uncinus produces no significant weather on the ground, although snow showers may be visible immediately below cloud level. These usually evaporate well before reaching the ground and are therefore classified as virga (see p. 220).

As cirrus uncinus generally indicates the presence of high-speed winds, pilots often associate this cloud with turbulence. In most cases, however, the turbulence would cause little discomfort to pilots or passengers.

Cirrus mares' tails are named for their resemblance to a horse's tail.

Cirrus Kelvin–Helmholtz

Appearing as a slender, horizontal spiral of cloud, cirrus Kelvin–Helmholtz is one of the most distinctive cloud formations. However, it tends to dissipate only a minute or two after forming and, as a result, is rarely observed.

The shape of this kind of cirrus is the result of a particular type of wind shear. In general, wind shear occurs when one layer of air slides across another layer moving at a different speed or in a different direction (or both). This gives rise to vertical eddies that produce a regular pattern of air waves (see p. 205).

In most cases, wind shear creates a series of gently undulating cloud formations along the tops of the waves. In the case of the Kelvin–Helmholtz formation, however, the eddies are more powerful, and carry the cloud over the peak and down the other side, so that the waves "break" in the manner of ocean waves breaking as they approach the shore. As these waves complete a circulation, they create a distinctive corkscrew pattern.

WEATHER WATCH

Worldwide

Above 16,500' (5,000 m)

Saturation of an air mass at high levels, combined with wind shear

None

Moderate to severe levels of turbulence at cloud level

This form of instability also occurs in fluids and in the Earth's outer atmosphere. It was first described in the late nineteenth century by Baron Kelvin (1824–1907), a Scottish physicist, and Hermann von Helmholtz (1821–94), a German physicist—hence the name of the cloud.

Kelvin–Helmholtz waves are probably quite common in the upper troposphere, but generally there is insufficient moisture present to generate cloud and render the pattern visible.

The presence of this cloud indicates a degree of wind shear that is likely to produce moderate to severe turbulence at cloud level. In the absence of cloud, the same process can be a major source of clear air turbulence at high levels. As this turbulence is invisible and does not show up on radar, aircraft may encounter it unexpectedly.

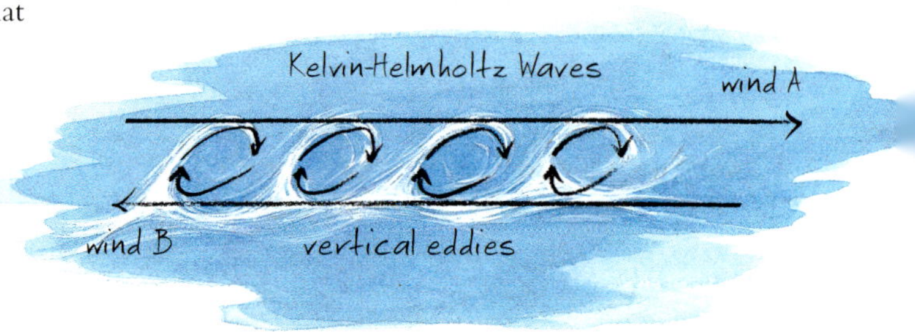

Cirrostratus

Cirrostratus is an even layer of cirrus that covers a wide area of the sky. As with other kinds of cirrus, it is formed when a moist air mass is lifted to a level where it cools to saturation and forms ice crystals. In the case of cirrostratus, this lifting occurs on a large scale.

Meteorologists distinguish various types of cirrostratus. Among the most common are fibratus and nebulosis. The former consists of long, thin filaments, known as striations, that spread out across a wide area of the sky. The even texture of this formation results from the ice crystals being blown by strong, steady, high-level winds.

In the case of cirrostratus nebulosis, the uplift that gives rise to the cloud is very gentle, and the resulting ice-crystal layer is extremely thin, with vague edges that are difficult to discern and a lack of the texture or fiber common to other cirrus clouds. Frequently, the only sign of cloud formation will be a slight diminution of the intensity of sunlight.

WEATHER WATCH

2

Worldwide

Above 16,500' (5,000 m)

Saturation of a large air mass at high levels

If the cloud is thickening, it may indicate the approach of a frontal system

When it forms in a very thin layer, cirrostratus nebulosis often gives rise to haloes, sun dogs, and iridescence (see pp. 259–61).

Occasionally, snow showers fall from cirrostratus formations, but these usually evaporate before reaching the ground and are classified as virga (see p. 220). If a build-up of cirrostratus from one direction is taking place, this is a sign of an increase in moisture at upper levels and may indicate the approach of a weather system such as a cold front.

Cirrostratus formations may cause slight turbulence at cloud level, but this is unlikely to affect aircraft operations or discomfort passengers.

Cirrostratus fibratus (above) consists of filaments of cloud. Cirrostratus nebulosis often creates optical effects (right).

Cirrocumulus

Cirrocumulus, like cirrostratus (see p. 215), occurs when a large area of moist air at a high level of the atmosphere reaches saturation and forms ice crystals. What differentiates cirrocumulus from cirrostratus is the presence of instability at cloud level. This gives the cloud its cumuliform appearance.

In isolation, this formation does not normally have any great significance. However, if there is a steady increase in this cloud over a period of time, it may indicate the approach of a frontal system.

WEATHER WATCH

Worldwide

Above 16,500' (5,000 m)

Saturation of a large air mass at high levels, combined with instability at cloud level

Increasing cover may indicate an approaching frontal system

Cirrocumulus is one of the most attractive of all clouds, often forming spectacular patterns that may stretch for hundreds of miles across the sky. One dramatic form of cirrocumulus is cirrocumulus undulatus, which appears as a fine, rippled pattern in the sky.

As with other undulatus forms, these ripples are produced by atmospheric waves generated by wind shear (see p. 205). However, in the case of cirrocumulus undulatus, the entire structure has a much finer appearance. This is partly due to the fact that atmospheric waves formed at high altitudes tend to have a shorter wavelength than those formed in the middle layers, but it is also a result of the greater distance between the cloud formation and the observer on the ground.

Cirrocumulus undulatus (above). A cirrocumulus formation, depicted by Charles Blunt in his 1849 book, The Beauty of the Heavens *(left).*

Contrails

This is an exceptional formation in that it does not occur naturally. Rather, it is an artificial cirrus cloud produced by high-level aircraft operations. The term contrails is short for condensation trails.

All aircraft engines emit water droplets from their exhausts. When an airplane flies through the upper levels of the troposphere, where temperatures are normally well below 32° F (0° C), these droplets immediately freeze to form ice crystals, creating an artificial cloud.

Often, the surrounding air mass will contain little moisture, and the resulting cloud will be thin, short-lived, and invisible to an observer on the ground. However, if the surrounding air mass is close to saturation, the cloud produced will be much broader and longer and may last for half an hour. It is this visible formation that is known as a contrail. For the weather-watcher, a long-lasting contrail can be a useful sign, as it reveals the presence of significant high-level moisture. This, in turn, may indicate the approach of a frontal system.

WEATHER WATCH

Worldwide

Above 16,500' (5,000 m)

Aircraft operations at high levels combined with significant moisture in the surrounding air

If long-lasting, can indicate an approaching frontal system

Contrails also have a military and strategic significance in that they reveal the presence and location of high-level aircraft that would normally be invisible to the naked eye. The enduring memory of many who witnessed the Battle of Britain, during the Second World War, is the criss-crossing spider's web of contrails that occurred day after day, high above England, as the Luftwaffe and the Royal Air Force engaged in combat in the upper levels of the troposphere.

Long-lasting contrails may indicate an approaching front.

Precipitation

Rain

Rain is defined as precipitation that reaches the ground in liquid form. Initially, rain forms in a cloud as either water droplets or ice crystals (see p. 46). These then grow large enough to fall from the cloud under the influence of gravity, the ice crystals melting on their way to the ground.

WEATHER WATCH

- Worldwide, except polar regions
- Water droplets or ice crystals falling from a cloud under the influence of gravity
- Increased humidity at ground level
- Prolonged downpours may cause flooding

Sometimes, water droplets or ice crystals fall from a cloud but evaporate in mid-air. This creates an effect that resembles a dark fringe hanging from the cloud base. This phenomenon is known as virga, and it occurs when there is a deep layer of dry air, or a shallow layer of extremely dry air, beneath the cloud.

Because virga does not reach the ground, it cannot be classified as precipitation. However, the evaporation that produces virga increases the water vapor content in the layer of dry air, making it increasingly likely that subsequent falls will reach the ground.

Classifying Rain

Rain that does reach the ground can be defined in a number of ways. In the United States, liquid precipitation is classified on the basis of the size of the raindrops and the extent of the associated visibility. Precipitation that consists of raindrops

An isolated shower falling from a congestus cloud.

that are less than $\frac{1}{50}$ inch (0.065 mm) in diameter and fall close together is defined as drizzle. Drizzle is categorized as light, moderate, or heavy depending on the visibility. Larger raindrops, or smaller drops that are widely separated, are classed as rain, which is defined as light, moderate, or heavy according to the amount that falls and the visibility.

Sudden Shower at Ohashi Bridge at Ataka, *by Ando Hiroshige (1797–1858).*

While this system is precise, the amateur weather-watcher is unlikely to make use of it. A simpler and more practical distinction, adopted in some other countries, defines the type of precipitation according to the cloud that produces it. Under this system, liquid precipitation is classed as either rain or showers. Rain refers to falls from stratiform clouds, particularly stratus and altostratus (see pp. 190, 204). These clouds normally cover a wide area, so rainfall from stratiform clouds tends to be widespread and relatively long-lasting.

Showers, on the other hand, refers to falls from cumuliform clouds (see pp. 196–203). These falls tend to be localized, and may only last for a minute or so. However, some showers may be heavy, particularly when associated with thunderstorms. Dry spells between showers normally last much longer than the shower itself. However, if there is a lot of cloud about, a number of showers may occur with only short dry spells between them.

Flooding may be caused by both rain from stratiform clouds and showers from cumuliform clouds (see p. 228). Persistent rain may lead to flooding over an extensive area, while heavy showers are more likely to cause flash floods.

Rain from stratus cloud (above). Showers from cumuliform clouds may be short-lived but heavy, particularly in the tropics (below).

Freezing Rain

In wintry conditions, when temperatures at cloud level are below zero, any water droplets that fall from clouds will be supercooled (see p. 40). This means that they are likely to freeze as soon as they encounter a colder layer of air or a surface whose temperature is below 32° F (0° C). Precipitation that freezes in either of these ways is known as freezing rain.

In the former case, the rain turns into tiny pellets of ice in mid-air. In the United States, this type of frozen rain is known as sleet. In Australia and the United Kingdom, it is referred to as ice pellets, while sleet is used to describe a fall of partially melted snow.

WEATHER WATCH

Common in regions that experience winter snows

Water droplets falling through subfreezing air or encountering frozen ground

Ice deposits at ground level

Slippery surfaces, ice accretion on aircraft and boats

Probably the most important distinction to be made between the various types of frozen precipitation is the difference between the ice pellet form of frozen rain and hail (see p. 226). The latter only forms in a thundercloud, while freezing rain may fall from any cloud that can produce rain, provided that the air is cold enough.

Surface Conditions

When large supercooled droplets strike subfreezing ground, they tend to spread out on impact before freezing, coating surfaces with a layer of clear ice known as glaze. This type of ice can produce hazardous conditions, making it extremely difficult to drive or even walk. A heavy downpour in these conditions is known as an ice storm. The accumulation of glaze on exposed objects as a result of an ice storm can cause significant structural damage—it has been known to bring down overhead wires and tree branches.

Supercooled water droplets from clouds may freeze as they fall through a deep layer of cold air, forming tiny pellets of ice.

Glaze on trees and grass (above). A close-up of glaze on flowers (left).

When very small supercooled droplets strike subfreezing ground, they tend to freeze immediately on impact, trapping air between them. This produces an opaque, granulated coating of ice, known as rime, which is not as slippery as glaze.

Ice pellets normally shatter on impact, scattering ice debris across the ground. However, if the ice pellets have not completely frozen through, water from inside the pellets may spread across the ground, forming a glaze as it freezes. If carried by a strong wind, ice pellets can sting exposed skin, causing great discomfort.

Once glaze has formed, it normally thaws in a few hours. However, there have been occasions when glaze has persisted for days. The most extreme case on record was during the winter of 1969, in Connecticut, in the United States, when glaze remained on trees for six weeks.

While freezing rain can be very inconvenient, the greatest associated hazard is ice accretion on aircraft and boats. If an airplane flies through a supercooled cloud, ice will quickly form on its body and wings, altering its speed and aerodynamics. A substantial build-up of ice on the masts of a sailing boat at sea can cause the vessel to capsize.

These supercooled droplets have frozen on impact with the ground.

Snow

A landscape draped in a thick mantle of fresh snow is one of nature's most magnificent sights. Snow is common during the winter months in Europe and North America and is a permanent feature of many mountaintops throughout the world. Mount Kilimanjaro, in Tanzania, maintains a permanent cap of snow despite being only 3 degrees south of the equator.

WEATHER WATCH

✦ *Common in high and middle latitudes but may occur near the equator at high altitudes*

◉ *Ice crystals bonding in cloud and then falling through cold air*

➤ *Low temperatures and build-up of snow at ground level*

⚠ *Heavy falls combined with high winds may lead to blizzards, whiteouts, and avalanches*

Snow begins as ice crystals that form a cloud when water vapor freezes around minute solid particles in the middle and upper levels of the atmosphere, where the temperatures are well below 32° F (0° C). The individual ice crystals gradually bond, forming snowflakes. Once the snowflakes are heavy enough, they fall to the ground.

Ice crystals form in a vast array of shapes depending on the temperature and humidity of the surrounding air mass. With the invention of the microscope, the beauty and diversity of ice crystals became apparent for the first time. An American farmer, William Bentley (1865–1931), photographed thousands of magnified ice crystals, and noted that, although there were identifiable crystal systems, no two crystals were identical. In order to study individual crystals, Bentley had to catch the flakes on a velvet-covered tray, tease the crystals apart with a probe, and smooth them out with a feather.

Many snow crystals occur as hexagonal formations, but others form shapes that resemble triangles, columns, or needles.

The Temperature of Snow

Often snow that falls from a cloud melts as it descends, and reaches the ground as rain. However, the melting process extracts latent heat from the surrounding air, causing the air temperature to cool and making it increasingly likely that subsequent snow will reach the ground.

Interestingly, the ideal conditions for snow are temperatures close to and just below zero,

Snow drapes sequoia trees in Sequoia National Park, California, USA (above). A winter snowfall in Colorado Springs, USA (left).

rather than colder temperatures. This is because the warmer the snow, the more moisture it will contain, and hence the bigger the flakes will be; and because a temperature close to 32° F (0° C) will allow snow to melt, refreeze, and combine in larger flakes. As a result, very slight changes in temperature can mean the difference between snow or rain, making accurate forecasts difficult.

Snow can settle on the ground in a variety of forms, depending on wind, temperature, and humidity. Air temperatures well below freezing produce small, powdery flakes that provide ideal conditions for skiing. Snowflakes that form in temperatures closer to 32° F (0° C) are larger and wetter and tend to stick to surfaces. Strong winds may pile up snow in hollows and against houses, in what are known as snowdrifts. Once snow has settled, it may melt and refreeze, becoming harder and more compacted in the process.

In mountainous areas, a build-up of snow can cause avalanches, which may sweep down steep slopes, destroying everything in their path. This is often the result of new falls of loose, powdery snow settling on a hard base formed by earlier falls.

More common, but just as dangerous, are blizzards. These are caused by a combination of heavy snow, low temperatures, and strong winds, and can bring cities to a standstill. An associated phenomenon is a whiteout, when heavy snow and low cloud render the ground and the sky indistinguishable, making navigation impossible.

In countries where these conditions occur, blizzard warnings are among the most important weather forecasts issued. Accurate forecasts can reduce fatalities by ensuring people are indoors by the time a blizzard occurs, and by alerting emergency services before conditions deteriorate.

Hail

Perhaps the most destructive form of precipitation is hail. These frozen lumps of ice are produced by thunderstorms and are responsible for injury and property damage worth millions of dollars around the world each year.

WEATHER WATCH

- *Widespread, most common in cooler middle latitudes*
- *Formation of lumps of ice in a thundercloud*
- *Thunder, lightning, and rain*
- *Large stones are a threat to aircraft, people, and property*

Hail is produced when supercooled water droplets (see p. 40) are circulated within the updraft area of a cumulonimbus cloud. As the droplets pass through areas of differing temperature and humidity, different types of ice build up on them. Where temperatures are just below freezing and there are many supercooled water droplets, layers of clear ice form. In colder parts of the cloud, where there are fewer and smaller water droplets, freezing is so rapid that bubbles of air are trapped, forming opaque rime ice. This layering effect is enhanced as the hailstone alternately melts and freezes as it encounters warmer and colder air.

Most hailstones are around the size of a pea; however, some grow as large as golf balls, or even oranges. The size and number of layers will depend on how long the hailstone remains in the storm—hailstones consisting of 25 ice layers have been recorded.

Hailstones (left) form as alternate layers of clear and opaque ice build up on a water droplet. This polarized cross-section of one of the largest recorded hailstones (below left), which fell on Kansas, USA, in 1970 and was as large as a grapefruit, clearly shows the layers.

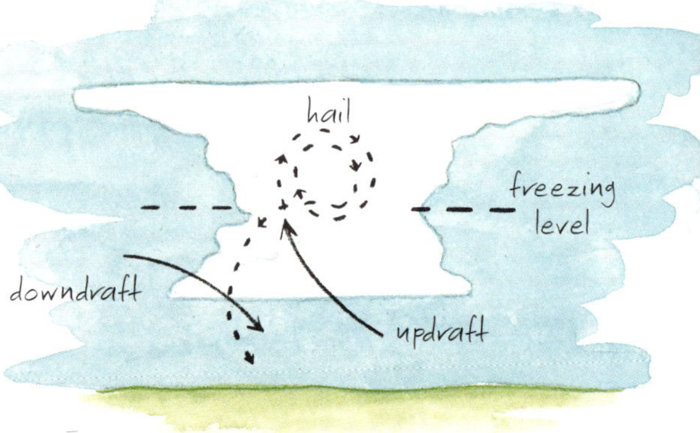

Hailfalls can be heavy but usually pass within a short time, because hail is produced only in the updraft area of a storm.

Hail will eventually fall from the cloud if it becomes too heavy to be supported by the updrafts, if the updrafts weaken, or if it is thrown clear of the updraft area.

Hail formation requires the strong updrafts associated with spring and summer storms. However, hail is unlikely to reach the ground in tropical areas as the high temperatures will melt the stones. Hail-producing storms are therefore most common around the middle latitudes, with parts of central North America reporting the highest incidence.

Some nineteenth-century farmers fired cannons into clouds to try to avert hailstorms.

Freak Hail

Throughout history, there are records of freak hailstorms, producing enormous stones. In 1888, in northern India, baseball-sized hail reportedly killed 250 people, as well as large numbers of livestock. More recently, in 1986, a hailstorm over Bangladesh produced 2¼ pound (1 kg) stones, which killed 92 people.

Hail causes a great deal of damage to property. Cars are particularly vulnerable and insurance claims run to hundreds of millions of dollars each year. Hailstorms also present a significant aviation hazard, although the advent of onboard radar has lessened this risk.

Signs of hail in an approaching storm include a greenish tinge at the base of the cloud, or a whitening of any rainfall. If hail looks likely, move people and pets indoors. Fortunately, hailstorms tend to be brief events, as they occur in only a small part of a moving storm.

Floods

Floods often involve property damage and loss of life, and are therefore classified as disasters. In fact, floods account for 40 percent of all deaths from natural disasters worldwide.

Yet in some parts of the world, floods are part of the natural annual weather cycle. For thousands of years, flooding along the Nile has sustained agriculture and hence civilization (see p. 62). Today, many seasonal tropical regions are dependent on annual flooding to nourish crops and provide water that can be stored for the dry months of the year (see p. 150).

WEATHER WATCH

- *Widespread over all continents except Antarctica*
- *Intense, localized downpours or heavy rainfall over an extended period*
- *Heavy rain, increased humidity, mudslides*
- *Can cause extensive property damage and loss of life*

Flash Flooding

A variety of meteorological conditions will cause flooding, but two basic types of flooding can be distinguished—flash floods and broadscale floods. Each of these has different causes, characteristics and timescales.

Flash flooding occurs when intense, short-term rainfall cannot be dispersed by soil absorption, runoff, or drainage. The most frequent cause of flash flooding is the slow-moving thunderstorm, which can deposit huge amounts of water over a small area in a very short time.

Floods can range from flash floods (above) to large-scale inundations, such as that described in the Book of Genesis in the Bible. In this Venetian fresco (left), Noah sends a dove in search of land.

Flooding at Tewkesbury, England, in 1995 (above). The aftermath of the Johnstown flash flood in Pennsylvania, USA, in 1889 (right).

Fast-moving storms are less of a problem in this respect, because they spread rain over a large area.

Flash floods often occur in valleys and gorges. When humid air is blown toward a mountain, it will rise and may develop into a storm, unleashing torrents of rain. If winds keep the storm stationary, water may rush off the slopes into the valley below. Gorges act like funnels, intensifying the flow until it is so powerful that it will destroy everything in its path. One of the worst flash floods in the United States occurred under such conditions. On 31 July 1976, a storm above Big Thompson Canyon, in the Rockies, filled the canyon, causing a flood that killed 139 people.

Surprisingly, slow-moving storms can also cause flash flooding in deserts. Hard, dry soil absorbs very little rain, so a sudden downpour can turn dry creek beds into raging torrents in a very short time. In fact, more people drown in North American deserts than die of thirst.

For similar reasons, flash flooding is increasingly common in cities. As more ground is covered with asphalt and concrete, less rain can be absorbed. Once drains overflow, water can flood rapidly along streets and alleys.

Sometimes, flash floods can be caused by the very structures that have been set up to control flooding. While dams and levees are often effective, they have been known to fail spectacularly, triggering devastating floods in the process. Another of North America's worst recorded flash floods occurred on 31 May 1889 in Johnstown, Pennsylvania, as a result of the collapse of a dam. Prolonged rain followed by a heavy downpour placed intense pressure on the South Fork dam, upstream from Johnstown. The 72 foot high (22 m) dam collapsed, sending its contents into the valley below. An hour later, the water hit Johnstown, carrying a wall of

Broadscale flooding will often cause rivers to overflow, demolishing bridges (above) and isolating livestock (left).

debris 30 feet (9 m) high. The impact destroyed much of the town, killing 2,209 people.

While flash floods can be devastating, they are usually short-lived, and disperse almost as quickly as they begin. Their effect also tends to be concentrated in a relatively small area. This is in marked contrast with broadscale flooding.

Broadscale Flooding

Broadscale flooding is normally associated with a frontal system, such as a cold front or low-pressure cell (see p. 34), that produces prolonged rain over an extensive area. Such flooding often begins along a river, which may break its banks and overflow into the surrounding countryside, gradually saturating the ground. Unlike a flash flood, this type of flooding may take several weeks to reach its peak.

The great Mississippi River flood of 1993—the longest and costliest flood in America this century—is a good example of broadscale flooding. The build-up began in December 1992 with heavy winter snowfalls in the Midwest. This snow melted rapidly during March and was followed by over 16 inches (41 cm) of rain in the upper Mississippi Valley during the month of April. Unseasonably heavy rainstorms then occurred along the Mississippi and Missouri rivers during June and July, causing both rivers to burst their banks in many places. Widespread flooding ensued, covering 16,000 square miles (41,000 km^2) in nine states.

In Chinese mythology, the dragon is the symbol of the seas and the rains. Offerings are made to dragon-kings in times of drought and in order to avert floods.

Hurricanes are a major cause of broadscale flooding over coastal and adjacent inland areas. These storm systems produce heavy rain, and are usually accompanied by a storm surge that can drive water far inland (see pp. 54, 250). As a hurricane weakens over land, it may settle as a depression, bringing further rain. Flooding associated with hurricanes is common worldwide. In the United States, in August 1969, Hurricane Camille swept into the Gulf of Mexico, producing devastating wind damage and widespread flooding along the Mississippi River. About half of the 256 people killed by the hurricane died as a result of the flooding.

Such destruction pales into insignificance when compared with similar disasters in parts of Asia. In Bangladesh, in 1991, a cyclone in the Bay of Bengal caused flooding that killed 150,000 people. Probably the most flood-prone river in the world is China's Yellow River, often referred to as "China's Sorrow". Major floods along this river have been recorded 1,500 times over the last 3,500 years, often with enormous loss of life. Flooding that occurred between 1887 and 1888 may have killed 2.5 million people, making it the most devastating flood in recorded history.

In such flood-prone areas, levee banks are often built along the sides of rivers. Some people have disputed their effectiveness, claiming that the damage that occurs when these systems fail is worse than that caused by normal flooding. Many critics of levees argue that a different approach to architecture, such as building houses on stilts, would be more efficient.

Much work has been done by weather services around the world to monitor rainfall and warn of flooding. However, flash floods in particular are likely to remain difficult to predict and prepare for.

The great flood of 1993 (above) affected 15 percent of the lower 48 states of the United States and caused US$20,000 million of damage. Monsoon rains in India (right) and Bangladesh frequently take a much higher toll in human terms.

Drought

While many weather phenomena are sudden and short-lived, drought is more insidious, gradually taking hold of an area and tightening its grip with time. In severe cases, it can last for many years, cover large parts of a continent, have a devastating effect on agriculture, and cause famine.

Contrary to popular opinion, drought is not merely low rainfall.

WEATHER WATCH

Occurs irregularly in most middle-latitude regions

Prolonged below-average rainfall over a wide area

Pronounced increase in aridity over affected area; dust storms and dust devils

Famine, damage to land and vegetation, starvation of livestock, severe dust storms, and wildfires

Rain is unevenly distributed around the world and some places will always receive less than others. Deserts, by definition, experience low rainfall. Tropical areas, too, have distinct wet and dry seasons, with little or no rain falling during the dry. Drought, therefore, is a relative term, based on the expected rainfall for an area at a given time of year. While droughts may be exacerbated by human activity, they are naturally occurring events and should always be planned for.

Precise definitions of drought vary greatly from country to country. In the United States, the term is generally used when an extensive area receives 30 percent or less of its normal rainfall over a minimum 21-day period. In Australia, a region is

Drought conditions in the Sahel region of Africa have persisted for over 30 years, resulting in thousands of deaths from famine.

The drought that occurred in Australia during 1982 and 1983 (above and right) may have been linked to changes in sea-surface temperatures in the eastern Pacific Ocean—the El Niño pattern.

said to be in drought if it receives less than 10 percent of its normal rainfall over a year, while in India, drought is declared if annual rainfall is less than 75 percent of the average.

A sophisticated system for measuring drought is the Palmer Drought Severity Index, developed by Wayne Palmer of the United States National Weather Service during the 1950s. This weighs up the balance between incoming water in the form of rainfall and stored moisture in the soil, and outgoing water in the form of evaporation from land and water surfaces, and the absorption and transpiration of soil moisture by plants. This information is then checked against climatic data and expressed as a negative or positive figure that represents the area's water level. For example, measurements between +2 and -2 signify normal conditions, while those between -2 and -3 indicate moderate drought; those between -3 and -4 indicate severe drought, and anything less than -4 signifies extreme drought.

A drought may break over a period of several months with the gradual return of normal rainfall. Alternatively, it may be broken suddenly by heavy rains, which then cause floods. The balance between drought and flood conditions that occurs in many parts of the world has led to a saying among meteorologists that "average rainfall is a drought plus a flood divided by two".

Where a prolonged drought occurs, the effects can be catastrophic. Water shortages can decimate crops and livestock, threatening the economic survival of farmers. Topsoil may become parched and dusty, and vegetation tinder dry, creating perfect conditions for dust storms (see p. 249) and wildfires (see p. 268).

In developing countries, drought can be even more serious, resulting in widespread famine.

The cyclical nature of drought may add to its impact, especially in arid and semi-arid regions. A prolonged period of above-average rainfall may give the inhabitants of such an area a dangerously optimistic view of the land's fertility. Nomads and farmers may extend their grazing areas and settle on previously uninhabitable land. When drought conditions inevitably recur, these people are ill-prepared for survival.

Such a scenario unfolded during the 1960s in the Sahel region of Africa. This area experiences extremely low rainfall as a rule, yet during the early 1960s there was a succession of exceptionally good seasons. As a result, settlers pushed farther into the desert. Then, at the end of the decade, a drought began that has continued virtually uninterrupted to the present day and led to the deaths of thousands of people in ensuing famines.

During droughts in Africa, animals, such as hippopotamuses (right), gather at water holes until rain (above) brings relief.

Dust Bowl Days

The worst drought in modern American history, in terms of persistence and loss of rural production, was the Dust Bowl drought of the 1930s. Vast areas of the Midwest suffered well below-average rainfall for almost a decade. By the mid-1930s, some 50 million acres (20 million ha) were severely drought affected. The previously prosperous Great Plains became a bleak and windswept "dust bowl", and huge agricultural losses were incurred. Relief finally arrived in 1940, when significant rains occurred across the region. After similar falls in 1941, the disaster was all but over.

Predicting Drought

Much work has been done over the last few decades to identify the causes of drought, and it now seems clear that prolonged periods of abnormally low rainfall over the Americas and Australia are linked to sea-surface temperature patterns across the Pacific and along the west coast of South America. These patterns can change rapidly, and this may explain the sudden breaking of droughts.

The best known of these patterns is the El Niño effect (see p. 104). This has probably caused trade winds from the eastern Pacific to weaken on occasion. Since these winds normally bring moisture to the western Pacific, El Niño may be responsible for periods of below-average rainfall in eastern Australia.

An unusually pronounced El Niño event in 1982–83 has been blamed for the severe drought experienced over much of Australia at that time. During 1982, the usual winter and spring rains failed to materialize over the eastern states, and by early 1983, extensive areas were reporting their lowest rainfall on record. The drought broke later that year, but not before dust storms and wildfires ravaged much of southeastern Australia. The total cost of the drought has been estimated at over US$2,200 million.

The linking of sea-surface temperatures with droughts is an exciting advance in long-term forecasting, and may eventually pave the way for accurate drought predictions. Governments, emergency services, and farmers could then take protective measures, such as water rationing, crop selection, and backburning of bushland well in advance of the onset of dry conditions.

Low water levels during a drought in Wales in 1989 (above) revealed old buildings that had been submerged when this reservoir was built. The Dust Bowl drought of the 1930s (above right) led to mass migrations within the United States.

Storms

Thunderstorms

The magnificent anvil-shaped cloud of the mature thunderstorm has long been an object of awe and fascination because of its capacity to unleash devastating rain, wind, hail, and even tornadoes, as well as awesome displays of lightning and thunder.

Thunderstorms occur under varied conditions, but are most common in spring and summer in tropical and subtropical zones. Air-mass storms (see p. 48) tend to occur in the late afternoon or evening when heating of the ground has reached its peak. Storms that result from frontal systems can occur at any time, but ground heating will tend to intensify their development.

WEATHER WATCH

Worldwide, except Antarctica; common in the tropics

2,000–35,000' (600–10,500 m)

Powerful convection assisted by atmospheric instability

Heavy rain or hail, strong winds

Lightning, wind, hail, and tornadoes; severe turbulence in cloud

Each day, approximately 40,000 thunderstorms occur throughout the world. The most thunderstorm-prone area is the southeastern United States, with some parts of Florida experiencing thunderstorms on an average of 100 days a year.

A typical thunderstorm may last up to two hours, although it will generally be at its mature stage for only about 15 to 30 minutes. After this, it will start to dissipate, often leaving only a few wisps of high-level cloud behind.

The distinctive pattern on the base of this summer storm indicates severe turbulence.

This statue of Zeus brandishing a thunderbolt dates from 460 BC. In Greek mythology, Zeus controlled the rain, clouds, and thunder.

STORM ALERT

There are a number of ways that the weather-watcher can tell if there are storms in the area. If the terrain is reasonably flat and the sky is not obscured by low-level stratus, a towering cumulonimbus cloud may be visible up to 200 miles (320 km) away. The direction in which it is moving can sometimes be determined by observing the shape of its anvil (see p. 200). Normally, there are long and short parts to the anvil, with the long section spreading out in the direction in which the upper-level winds are blowing. This is generally the best indication of the movement of the storm. Surface wind is not a good indicator because thunderstorms are affected by the speed and direction of the wind at all levels of the troposphere.

In some cases, when the sky is covered with many different types of cloud, or the terrain is mountainous, a thundercloud may be hidden from view. However, storms up to 100 miles (160 km) away can be detected by using a radio receiver. Tune the receiver into an area on the dial where no transmissions are taking place, and then turn up the volume. If there is an active thunderstorm around, you will hear distinctive bursts of static, produced by the storm's lightning. An increase in the volume of the static indicates that the storm is getting closer.

If an active thunderstorm is less than about 20 miles (32 km) away, you should be able to hear it. Because light and sound travel at different speeds, you can approximate the storm's distance by counting the interval between a flash of lightning and the associated sound of thunder. As a rough guide, every 5 second interval is equal to 1 mile (3 seconds is equal to 1 km) between you and the storm. If the interval between the lightning flash and the thunder decreases, the storm is getting closer, with simultaneous lightning and thunder indicating that you are directly beneath it.

Cloud-to-ground Lightning

Lightning occurs when there is an electrical discharge within, or around, a thunderstorm (see p. 50). Cloud-to-ground lightning occurs when the electrical charge travels between a negatively charged cloud base and the positively charged ground. This is the most spectacular variation of lightning, forming brilliant, jagged bolts between the sky and the ground.

Each lightning stroke lasts a fraction of a second. Sometimes a number of strokes is needed to discharge the electrical build-up, giving the lightning a flickering appearance. Often the main stroke combines with smaller offshoots that discharge into the air or inside the cloud.

WEATHER WATCH

Worldwide, except Antarctica; common in the tropics

Usually from cloud base to ground; rarely from cloud top

Electrical discharge between cloud and ground

Heavy rain or hail, strong winds from associated thundercloud

Strikes may damage property and cause serious injury or death

The ground, and almost any solid object in connection with the ground, will conduct electricity more effectively than air. This means that elevated landmasses and tall objects such as buildings and trees are prone to strikes.

Most cloud-to-ground lightning occurs from the base of a cloud. However, a rarer form, known as a positive flash, occurs when positive charges higher up in a cloud react with negative ones on the ground, sending a mighty lightning bolt from the top of the cloud to the ground. Since the path of this type of stroke is much longer, the charge has to be far more powerful.

The color of lightning indicates the content of the surrounding air. The flash will appear red if there is rain in the cloud, and blue if there is hail. The presence of a significant amount of dust in the atmosphere will produce yellow lightning. White lightning indicates low humidity; as a result, this is the form of lightning most likely to generate fires on the ground.

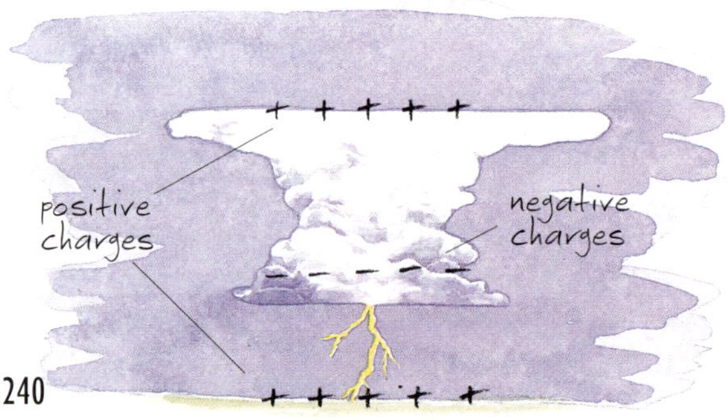

Striking Out

Although only about 20 percent of lightning reaches the ground, strikes occur somewhere on Earth over 100 times every second. In North America, about 400 people are struck by lightning each year, and about one in four of these strikes is fatal.

There are a number of precautions you can take to minimize your chances of being hit by lightning during a storm. If possible, move indoors. When lightning strikes a building, it tends to run along plumbing and electrical circuits, so you should avoid touching metal pipes or using any electrical equipment, including telephones and computers. One of the safest places to be is inside a car, as the car's tyres provide insulation. Aircraft are also safe, because they are not in contact with the ground and therefore cannot conduct electricity.

If you are caught outdoors, do not shelter beneath isolated trees, as they are favorite pathways for the lightning's leader strokes. Keep clear of metal objects such as wire fences, which can conduct electricity over considerable distances.

White cloud-to-ground lightning indicates an absence of moisture in the air. It is therefore the form of lightning most likely to cause fires.

Should your hair begin to stand on end, this may mean that you are within the area of positive charge below the cloud and that a strike is imminent. If this happens, crouch on all fours at once and keep your head low. Do not lie full-length on the ground, as this will increase your contact with any charges that may be conducted through the ground by wet soil.

If someone is struck by lightning, expert medical attention should be requested at once and cardiopulmonary resuscitation attempted.

The greatest myth associated with cloud-to-ground lightning is that it never strikes the same place twice. The top of the Empire State Building is struck about 500 times a year and was once struck 15 times in just 15 minutes.

Cloud-to-cloud Lightning

Cloud-to-cloud lightning is the most common form of lightning. Most often it occurs within a cloud, and involves electricity passing between the negatively charged base of the cloud and its positively charged upper levels. This internal lightning stroke often illuminates the cloud from within. A large flash can produce a spectacular snapshot of an entire cumulonimbus, which may remain visible for up to half a second if there is a succession of strokes up and down the leader path (see p. 50).

WEATHER WATCH

Worldwide, except Antarctica; common in the tropics

Anywhere within the height range of a cumulonimbus cloud

Electrical discharge between one thundercloud and another, or within a single thundercloud

Heavy rain or hail, strong winds from associated thundercloud

Less frequently, cloud-to-cloud lightning involves an electrical discharge between opposite charges in two adjacent clouds. This will normally occur between the positively charged top of one cloud and the negatively charged base of the other.

Because cloud-to-cloud lightning normally occurs at higher altitudes than cloud-to-ground lightning (see p. 240), it may be seen from some distance away, particularly at night. Indeed, a large cumulonimbus cloud will be visible up to 200 miles (320 km) away if the surrounding terrain is reasonably flat.

Thunder is usually audible only up to around 20 miles (32 km) from the lightning stroke that created it. This means that cloud-to-cloud lightning often appears to the observer as a "silent storm", with frequent flashes illuminating the sky amid eerily silent surroundings.

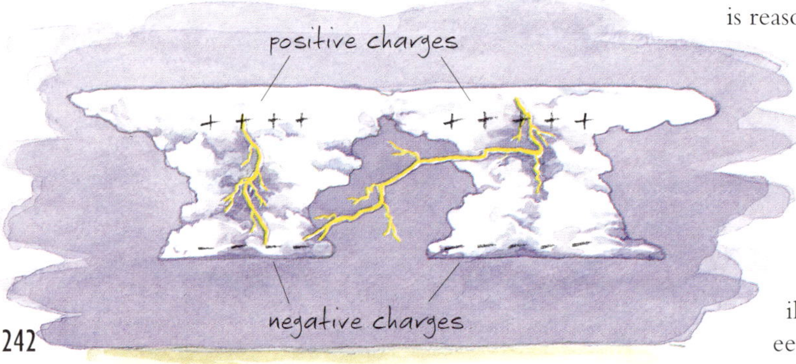

Cloud-to-air Lightning

This form of lightning occurs when an electrical discharge takes place between a build-up of one type of charge within a cumulonimbus cloud and an area of opposite charge in the surrounding atmosphere. This type of lightning tends not to be as powerful as cloud-to-ground lightning (see p. 240), and one stroke is normally enough to reduce the difference in charges to below critical levels. As a result, repeated flashes along the same cloud-to-air leader stroke are unusual.

Cloud-to-air lightning normally occurs between the air and the positively charged upper regions of clouds. It does also occur in the lower parts of clouds, but usually in combination with the positive flash form of cloud-to-ground lightning, when the cloud-to-air bolts will appear as weaker offshoots from the main flash. This happens because, in the lower layers, the difference in charges between the cloud and the ground is normally greater than the difference in charges between the cloud and the surrounding air.

WEATHER WATCH

Worldwide, except Antarctica; common in the tropics

Anywhere within the height range of a cumulonimbus cloud

Electrical discharge between a thundercloud and the adjacent atmosphere

Heavy rain or hail, strong winds from associated thundercloud

Because cloud-to-air lightning normally occurs near the top of a cumulonimbus cloud, it is often seen from a considerable distance away. If the storm is too far away for the thunder to be heard—usually over 20 miles (32 km)—the observer may witness a "silent storm" (see p. 242).

When cloud-to-air or cloud-to-cloud lightning is obscured by clouds, the viewer may see only its flickering reflection in the adjacent clouds. Commonly known as "sheet" lightning, this is actually a simple optical effect rather than another type of lightning.

Tornadoes

A tornado is a violently spinning vortex of air that extends from the base of a storm cloud to the ground (see p. 52). It is associated with severe storm activity and is one of nature's most destructive phenomena, capable of generating winds of up to 300 miles per hour (483 kph) in extreme cases.

Exceptional tornadoes may last for hours and travel hundreds of miles. In the United States, the Mattoon–Charleston Tornado of 26 May 1917 covered 293 miles (471 km) in just under seven and a half hours. Most tornadoes, however, are far weaker than this, with some lasting only seconds and generating winds of less than 50 miles per hour (80 kph).

Tornadoes may occur as isolated incidents or in great numbers. In the United States, the so-called "Super Outbreak" of 3–4 April 1974 saw 148 individual tornadoes devastate an area from Alabama to Michigan.

WEATHER WATCH

- *Wherever thunderstorms occur; most common over Great Plains of North America*
- *From base of cumulonimbus cloud to ground level*
- *Rapid rotation of updrafts within a thunderstorm*
- *Destructive surface winds*
- *Threat to life and property*

Storms severe enough to generate tornadoes are most likely to occur in the middle latitudes. The United States is by far the most tornado-prone country in the world, enduring around 750 tornadoes annually. Most of these are confined to the Great Plains, with central Oklahoma having the dubious distinction of experiencing more tornadoes per acre than any

This 1896 lithograph by Kurz and Allison depicts the St Louis tornado of that year.

The color of a tornado depends on the dirt and debris it collects.

other location on Earth. Tornadoes also occur regularly in Australia, and occasionally in other countries such as the United Kingdom.

Tornadoes can occur at any time of the year. In the United States, there is an overall peak of activity in May and June on the Great Plains. However, tornadoes occur in different parts of the country at different times of year, with the focus of activity shifting from the Gulf in late winter to the Great Plains in midsummer, and moving southward again in fall.

Tornado Alert

There are two signs to look for when assessing whether a storm is severe enough to generate tornadoes. The first is the "overshoot" phenomenon, where the normally flat top of the storm's anvil displays an ominous bulge (see p. 48). This indicates that the upward rush of air near the center of the storm is so powerful that it has "punched" through the tropopause, bubbling up into the stratosphere. The second feature is an extensive and well-defined mammatus formation (see p. 202).

A tornado's movement can be erratic, creating a cycloidal damage path (like the track a spinning top takes on a flat surface). This explains why a tornado can demolish houses either side of one that is left untouched. While this makes it difficult to tell whether you will be caught in an approaching tornado's path, there are some precautions you can take. If possible, shelter indoors, particularly in a bathroom, because this is often the strongest room in the house. If caught outside, try to shelter in a ditch.

Waterspouts

Waterspouts are rapidly rotating columns of air that form over lakes and oceans. They resemble tornadoes over water, and in some cases, are exactly that. However, they generally do not require severe thunderstorms to generate and sustain their motion, and are more often associated with congestus clouds (see p. 196).

> **WEATHER WATCH**
>
> *Most frequent over coastal areas in middle latitudes*
>
> *Typically 1,500' (450 m), but can be as high as 3,000' (900 m)*
>
> *Tornado over water, or rotation of air mass over water*
>
> *Showers from associated cloud*
>
> *Can be a hazard to shipping*

Waterspouts are classified as either tornadic or non-tornadic. Tornadic waterspouts are formed by the same mechanisms as tornadoes on land (see p. 52), and are relatively rare. Less intense, non-tornadic waterspouts appear to be caused by a pre-existing rotation near the surface of the water, combined with some form of updraft. This produces a funnel of rotating air that extends from the water to the base of the cloud.

Non-tornadic waterspouts are most common in late summer or early fall. At this time the combination of warm sea-surface temperatures and cold air currents produces instability (see p. 42) and strong updrafts.

People often assume that a waterspout draws water up from the sea or lake below. In fact, apart from a small area of spray at the base of the spout, the water in the funnel is a result of condensation caused by very low pressure within the spiraling air mass.

Waterspouts can occur in isolation or in clusters. Sometimes, the first indication of a developing waterspout is a shadow on the water where the rotating air is disturbing the surface. Once fully formed, a waterspout will tend to move slowly along a curved path for about 15 minutes until cooler air gradually enters the funnel, causing the spout to rapidly dissipate.

Dust Devils

A dust devil is an upward-spiraling, dust-filled vortex of air that may vary in height from only a few feet to over 1,000 feet (300 m). Dust devils occur mainly in desert and semi-arid areas, where the ground is dry and high surface temperatures produce strong updrafts.

Dust devils resemble mini-tornadoes (see p. 52), but are generally nowhere near as intense or damaging. They normally begin when winds blowing around local terrain features create a rotating air mass in the low or middle levels of the troposphere. This rotation then combines with strong updrafts produced by surface heating of the ground to create a powerful, rising funnel of air. As the air rises, it draws up large quantities of dust. It is this dust that renders the funnel visible.

On some occasions, a cumulus cloud will form over the updraft area. This may give the appearance that the dust devil is emanating from the cloud, but this is never the case; it is the rotation in the surrounding air mass that generates the funnel.

The presence of the cloud is, however, an indication that the initial updraft was relatively powerful, and the most intense dust devils are often associated with a cumulus cloud. Particularly powerful devils have been known to rip the roof off a house or flip over a car, but normally dust devils pose little threat to life or property.

WEATHER WATCH

Arid areas, particularly during hot weather

0–1,000' (0–300 m) high

Lifting of dust by air-mass rotation combined with updrafts

Strong local winds, reduced visibility

Can cause damage to property

Dust devils are most common in desert and semi-arid areas.

Microbursts

A microburst is a brief, powerful gust of wind that appears to radiate from a central point on the ground. It is caused by strong downdrafts that form in the central part of a congestus or cumulonimbus cloud.

There are two distinct types of microburst—dry and wet. The dry microburst occurs in dry conditions, when a column of rain falls into a layer of dry air beneath the cloud, and immediately begins to evaporate. Since evaporation produces cooling, this accelerates the downward motion of the air column, producing a powerful gust of wind that spreads in all directions. Where there is warm air near the ground, this will tend to rise and counter the downdraft. However, the descending air may still reach the surface with some velocity. Because the precipitation usually evaporates completely, the only visible sign of a dry microburst will be raised dust.

The wet microburst is usually associated with heavy rain and, again, evaporation is the vital ingredient producing strong surface winds. However, in this case, the precipitation reaches the land below. Often, the wind and rain meet the ground with such force that they spread outward and upward, forming a distinctive curl.

Microbursts are a major hazard for aviation as they can destabilize an aircraft that is taking off or coming in to land. This has been the cause of a number of serious accidents in the industry.

WEATHER WATCH

Dry form common in arid areas, wet form common in wet climates

Cloud base around 5,000' (1,500 m) for dry; 2,500' (750 m) for wet

Powerful downdrafts

Brief, often powerful gusts of wind and heavy showers

Extreme aviation hazard

A wet microburst (above).

Dust Storms

Strong winds are always capable of lifting topsoil and scattering it over large areas, but occasionally, certain conditions combine to produce huge walls of moving dust that carry thousands of tons of soil and debris to another location.

Such events tend to occur after an extended drought has left the ground dry and dusty. If a vigorous cold front then moves across such an area, the ascending air at the face of the cold front may lift the topsoil, forming a huge moving wall of dust. This wall will be carried along by the front, with more soil feeding into the system as it advances.

A cloud of raised dust is generally considered to be a dust storm if visibility is reduced to about half a mile (1 km). It is considered severe if visibility decreases to a quarter of a mile (0.5 km) or less.

Dust can be lifted up as high as 10,000 feet (3,000 m), and travel several thousand miles, remaining airborne for days. Dust storms generated by vigorous fronts over southeastern Australia have carried soil right across the Tasman Sea to New Zealand, producing dust-colored "red snow" on the New Zealand Alps. A similar phenomenon occurs in North America, when dust storms from the Plains produce dust-colored snow and rain along the Atlantic coast.

Dust storms may be preceded by dust devils (see p. 247) that have detached themselves from the main front, but these are unlikely to cause much damage. Large dust storms frequently leave behind an enormous amount of fine dust that infiltrates every corner in a house, even making its way between book pages. By far the most serious damage caused by dust storms is the removal of valuable topsoil from farmland and other areas.

WEATHER WATCH

Most frequent in arid areas or temperate zones after drought

From the surface up to around 10,000' (3,050 m)

Lifting of dry topsoil by an active cold front

Strong winds

Restricted visibility; destruction of crops and farmland

A dust storm hits Alice Springs, Australia (top). A nineteenth-century illustration of a dust storm in Africa (above left).

Hurricanes

Little else in nature has the destructive force of a fully fledged hurricane. These massively powerful systems can produce sustained winds of 150 miles per hour (250 kph) with gusts of up to 190 miles per hour (300 kph), as well as intense bursts of rain and ocean surges that cause extensive flooding (see p. 54). While tornadoes can produce even stronger winds, they rarely last for more than a few hours. A hurricane, on the other hand, can last for weeks and cover thousands of miles.

A mature hurricane consists of bands of thunderclouds spiraling around the eye—a clear, almost calm area at the center of the storm. The whole storm system may contain hundreds of thunderstorms and measure up to 600 miles (970 km) in diameter.

To qualify as a hurricane, a storm must produce winds of over 74 miles per hour (119 kph).

WEATHER WATCH

- Between 5 degrees and 30 degrees North and South
- To around 60,000' (18,000 m)
- Intense convection over warm tropical oceans, combined with instability
- Destructive winds and rain
- Extreme threat to life and property

In 1992, the 145 mile-per-hour (235 kph) winds and 17 foot (5 m) waves caused by Hurricane Andrew (above) created US$25 billion dollars worth of damage. A repetition of this disaster was avoided in September of the following year when Hurricane Emily (left) veered away just before it struck the East Coast.

Hurricane Bonnie, as seen from the space shuttle (above). The aftermath of a hurricane on Kauai in the Hawaiian Islands (right).

In the Northern Hemisphere, rotating systems with lower wind speeds are known as tropical storms or tropical depressions.

In the western Pacific and China Sea area, hurricanes are known as typhoons, from the Cantonese *tai-fung*, meaning great wind. In Australasia and countries around the Indian Ocean, the same storm systems are known as tropical cyclones.

The clusters of storms that produce hurricanes occur only where sea temperatures are at least 80° F (27° C). This means that they usually originate in the tropics. To develop its distinctive rotation, the system must be at least 5 degrees from the equator, because this is where the Coriolis effect begins to have an influence (see p. 31).

Once spinning, a storm system tends to move farther away from the equator, although it is unlikely to continue beyond 30 degrees north or south. If a hurricane returns toward the equator, it usually begins to weaken. It is impossible for a hurricane to cross the equator, because there the Coriolis effect has no impact and the hurricane will lose all its rotational energy and decay into a cluster of thunderstorms once again.

Hurricane Folklore

Destructive hurricanes have been recorded many times throughout history, with accounts of cities being destroyed and shipping fleets sunk by furious winds and mountainous seas. The naval fleets of the Mongol emperor, Kublai Khan, were scattered by typhoons in 1274 and again in 1281, while preparing to attack Japan.

A contemporary illustration of a devastating hurricane that hit Georgia in the United States in 1940.

The Japanese, believing that these storms had been sent to protect their country from invasion, referred to them as divine winds, or *kami-kaze*.

Until recently, there was no sure way of knowing if a hurricane was approaching the coast, but often the sea provided some clues.

A hurricane at sea produces a swell that spreads from the center of the system and often runs well ahead of the storm. Thus, if a large swell is observed, particularly when there is little local wind, this may be a sign of a hurricane. From a high vantage point, it should be possible to observe the direction from which the swell is approaching, and from this, deduce the direction in which the hurricane lies. The farther the waves are from the storm, the greater the distance between them; so, if swell waves are breaking in increasingly rapid succession as the day progresses, a hurricane may be approaching.

Modern Surveillance

The development of satellite photography has allowed people to fully appreciate the spectacle and majesty of these revolving spirals of cloud and permitted scientists to monitor their movements closely. Satellite images now form the basis for hurricane forecasting around the world and, as meteorologists have become more expert at interpreting them, hurricane forecasting has continued to improve.

During the local hurricane season, meteorologists receive satellite photographs hourly. They look for large clusters of thunderstorm clouds over tropical oceans, which may be embryonic hurricanes. Once such a cluster is identified, scientists monitor sequences of satellite images for any sign of rotation.

If the satellite imagery reveals a rotating system moving into higher latitudes, the

In order to study hurricanes, research scientists undertake flights into the eye of the storm (above left). The towering ramparts of cumulonimbus clouds that surround the eye may measure up to 60,000 feet (18,000 m) high. The storm surge (right) normally accounts for the majority of deaths caused by hurricanes and a high proportion of property damage (above).

meteorologist looks for evidence of eye formation. Once this has been identified, and the surface wind speed has reached 74 miles per hour (119 kph), the weather service in question will then declare it to be a hurricane, and it will be officially named from the list provided by the World Meteorological Organization (see p. 55). A continuous watch is then kept on the hurricane, with warnings disseminated to shipping, aircraft, and the general public as long as the storm persists.

If a hurricane moves to within about 150 miles (240km) of the coastline, it is then within radar range, and can be accurately tracked. However, even at this stage, there is an element of uncertainty as hurricanes behave unpredictably. It may continue in a straight line, or it may stall or veer. The reasons for this erratic movement are the subject of continuing research.

The hurricane poses the greatest danger to human life once it reaches the coastline. Loss of life can be minimized by evacuating an area, but destruction of property is unavoidable. The surge of sea water associated with a hurricane can inundate large areas of coastal land, while high winds and further flooding from rain may cause additional damage.

As the hurricane crosses the coastline and moves away from the sea—its source of energy and moisture—it begins to dissipate rapidly, although rains may continue for a few days.

Optical Effects

Rainbows

Both dazzling and ephemeral, the rainbow has fired imaginations throughout time, often assuming great religious significance. However, it was not until the late 1660s that a scientific explanation for this phenomenon was provided.

In a classic experiment, Isaac Newton showed that when a beam of light passed through a glass prism it was refracted, breaking down into a spectrum of colors. From this, he inferred that white light was in fact a combination of all the colors of the visible spectrum (see p. 56).

This experiment suggested an explanation for the rainbow: raindrops act as millions of tiny prisms, breaking up sunlight into its component colors.

When light hits a raindrop, most of it passes straight through, but light at the edges of the raindrop is refracted into the colors of the spectrum and then reflected off the back of the raindrop. It is then refracted again as it leaves. This refracted light leaves the raindrop at an angle of about 42 degrees from the incoming rays. Each color emerges at a slightly different angle, depending on its wavelength.

Only one color will be visible from any one raindrop at a time, depending on the angle from which it is observed. The observer on the ground sees the combined refraction and reflection of light from millions of drops forming distinct bands of color: red, with the longest wavelength, is on the outside, ranging to violet on the inside.

A rainbow depends on the movement of the raindrops, and the position of the Sun and the observer. Because of these variables, no two people will ever see precisely the same rainbow.

WEATHER WATCH

Widespread, mainly during morning and afternoon

Up to 42 degrees above the horizon

Refraction and reflection of sunlight by raindrops

Alternate showers and clear sky

A secondary rainbow sometimes appears above the primary bow when light is reflected twice off the back of the raindrops.

A rainbow over the Sierra Nevada, California, USA (above). Fogbows are normally colorless or almost colorless (left).

Watching Rainbows

Since rainbows are formed by the interaction of sunlight and rain, they normally occur during showery weather. To see the rainbow, the observer must be between the Sun and the shower. The brightest rainbows occur when the raindrops are large, because these disperse light better.

A rainbow forms its greatest arc when the Sun is close to the horizon. The higher the Sun, the flatter the rainbow will be, until the Sun rises higher than 42 degrees above the horizon, when the rainbow will disappear altogether. If there were no horizon, the rainbow would appear as a full circle, as is sometimes observed from airplanes.

As rainbows depend on the Sun being close to the horizon, they are more common in the mornings and afternoons than in the middle of the day. This also means that they are more common in winter than in summer, and in middle and high latitudes than in the tropics.

In addition to rainbows, there are also moonbows and fogbows. A moonbow occurs when light reflected by the Moon refracts into visible colors on contact with rain. The resulting colors tend to be faint at best. Fogbows occur when sunlight strikes the water droplets in a fog. These bows are generally colorless, or almost colorless, because the tiny water droplets in fog don't disperse light well.

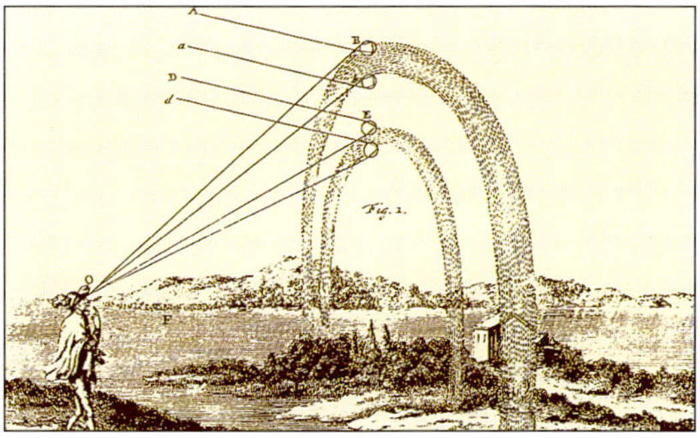

This seventeenth-century illustration shows how sunlight is refracted and reflected to form primary and secondary rainbows.

Coronas

WEATHER WATCH
Widespread but rare; most frequent in winter over mountains
Usually associated with middle-level clouds
Diffraction of light from Sun or Moon by a thin layer of middle-level cloud of uniform droplet size
Fair conditions associated with thin altostratus

A corona (meaning crown) describes one or more luminous disks of light that occasionally appear around the Sun or Moon. They occur when these objects are viewed through a thin layer of cloud consisting of water droplets.

A corona is caused by light diffraction—the slight bending of light as it passes by the edge of an object. This process causes the colors that make up white light to separate, because each color has a different wavelength and bends at a different angle. A corona forms when sunlight passing through a cloud is diffracted by water droplets in the cloud. The short wavelength of blue light diffracts the most and forms the inner ring, while red forms the outer ring. While orange, yellow, and green may be visible in bright coronas, blue and red are the predominant colors. Sometimes, several rings will be visible, becoming fainter the farther they are from the center.

A corona is best seen when the Moon, rather than the Sun, is providing the light, as the brightness of the Sun tends to blind the observer to the subtle effects of diffracted light. Small water droplets produce the largest coronas, and those of a uniform size produce the brightest colors. If the cloud contains a variety of differently sized droplets, the rings become deformed and irregular and the colors diffuse. Newly formed clouds tend to contain uniformly small droplets (see p. 42), so the best conditions for a colorful and well-defined corona are a bright Moon shining through a thin layer of newly formed, uniform cloud, such as altostratus (see p. 204).

 Coronas around the Sun (above) and the Moon (right).

Iridescence

Iridescence appears as irregular patches of color in middle-level clouds adjacent to the Sun or the Moon. It is best understood as a partial, or imperfect, corona (see p. 258), as it is formed by the same process of light diffraction around water droplets.

WEATHER WATCH

Widespread but rare; most frequent in winter over mountains

Usually associated with middle-level clouds

Diffraction of light from the Sun or Moon by a thin layer of middle-level cloud of uniform droplet size

Fair conditions associated with thin altostratus or altocumulus

Iridescence lacks the symmetry of a corona, appearing as diffuse patches of color within the cloud, or as fringes of color on the cloud's edge. The observer on the ground will see iridescence instead of a corona when the water-droplet cloud is too small to produce the symmetrical rings of a corona, or if the Sun or Moon is not directly behind the cloud.

The observable colors of iridescence depend upon the size of the water droplets in the cloud and the angle at which the observer sees them. Blue, which forms the inner ring of a corona, is normally the dominant color, but reds and greens may also be visible. The brightness of these colors increases with the number of droplets in the cloud and the uniformity of their size. As with coronas, uniformly small droplets produce the best visual effect, and therefore newly formed altostratus or altocumulus clouds (see pp. 204, 206) provide the best conditions for iridescence. Iridescence associated with the Sun will produce stronger colors, but they will often be overwhelmed by the brightness of the Sun. Moonlight produces paler colors but a better opportunity to see them.

Iridescence, although an unusual phenomenon, can occur in virtually any part of the world, although it is most common in winter over mountains. When observed over a large city, it is likely to generate great curiosity.

Iridescent clouds: most of the colors of the spectrum are visible.

Haloes

A halo is a white or faintly colored ring that sometimes appears around the Sun or (less often) the Moon. Whereas coronas (see p. 258) almost appear to emanate from these light sources, haloes form a thin, wide circle around them. The main difference between the two phenomena is that coronas require water droplets to form, while haloes require ice crystals. These crystals may be free-falling or within thin layers of cirrus cloud.

Normally, when sunlight or moonlight strikes ice crystals, most of the light is reflected, producing a completely white halo. However, if the incoming light strikes the falling ice crystals at a particular angle, some light may be refracted (see p. 56). In such a case, the halo will be faintly colored, with red appearing nearest the center and blue on the outside. This is the reverse order of the corona's colors, because of the difference between light refraction and diffraction (see p. 258).

WEATHER WATCH

Worldwide, most common in high latitudes

Often associated with cirrus clouds

Reflection and refraction of light by ice crystals

Cirrus cloud may indicate deteriorating weather conditions

A 22 degree lunar halo over the Arctic.

Most ice crystals are hexagonal, or six-sided, and the most common angle of refraction through such a crystal is about 22 degrees. The most common haloes are those produced in this way, and are known as 22 degree haloes. Differently shaped crystals, or those at a different angle to the Sun, may produce haloes of various sizes and shapes. Smaller haloes at 9 degrees and larger ones at 46 degrees are not uncommon, and sometimes a halo will only partially form, appearing as an arc.

In folkore, haloes have long been associated with approaching rain, and there is a grain of truth in this. The cirrus cloud that can produce a halo may well indicate an approaching frontal system. However, in many cases, this front will be inactive, or may veer away from the area, producing no rain. As a result, the forecasting value of haloes is limited.

Sun Dogs

Sun dogs, also known as parhelia or mock suns, appear as two bright spots on either side of the Sun, creating the eerie illusion of three suns in the sky. They often appear in conjunction with a 22 degree halo, and are produced under the same conditions.

Sun dogs are the result of sunlight passing through a thin layer of ice crystals, either contained within cirrus cloud or falling at lower levels. Sun dogs will occur only when the hexagonal ice crystals are oriented horizontally—that is, with the large, flat sides facing down—so a large number of randomly falling ice crystals is necessary to sustain sun dogs.

WEATHER WATCH

Worldwide, most common in high latitudes

Often associated with cirrus clouds

Reflection and refraction of light by hexagonal ice crystals

Cirrus cloud may indicate deteriorating weather conditions

In very cold climates, such as Antarctica (above), ice crystals forming close to the ground may produce sun dogs.

When well developed, the two sun dogs may appear delicately tinted with red on the inside and blue on the outside. Sometimes, only one sun dog will appear, or one will be considerably brighter than the other. More often, they will appear just as two bright or luminous areas on either side of the Sun, with no distinguishable color. It is possible for an unusually bright moon to create the same effect, but "moon dogs", as they are known, are very rare.

Sometimes, sun dogs will gain height with the Sun during the day, although this can continue only up to an angle of 45 degrees above the horizon. Once the Sun climbs beyond this level, the refracted light will not be visible to the observer on the ground.

If cirrus cloud is creating the sun dogs, it may indicate the approach of a frontal system or a developing low-pressure cell, but generally this is not a reliable method of forecasting.

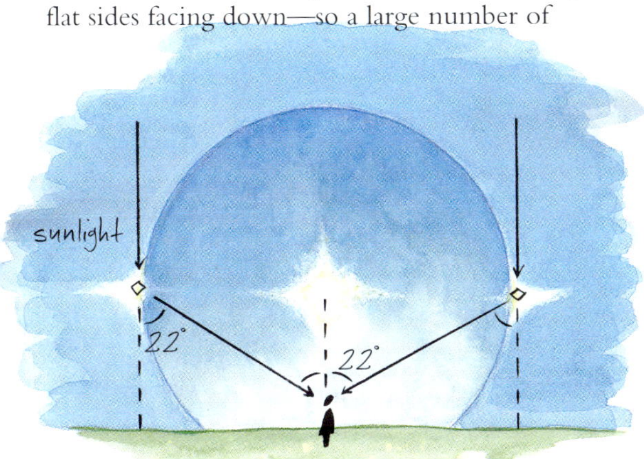

Auroras

One of nature's most stunning displays is the aurora, with its silently shifting curtains of luminous color often covering much of the night sky. Sometimes, it will appear as a colorful, glowing arc, remaining virtually stationary for extended periods. At other times, it will appear to glide eerily about the sky in a flowing waterfall of color.

WEATHER WATCH

Common in high latitudes

50–600 miles (80–1,000 km) above the Earth

Electron bombardment of the atmosphere by solar emissions

Possible links to unusual weather patterns

Can disrupt communications systems and cause power blackouts

Auroras occur in both hemispheres, and are known as either the aurora borealis (northern lights) or the aurora australis (southern lights). The word aurora is derived from the name of the Roman goddess of the dawn, and references to auroras exist in the writings of the ancient Greeks and Romans and appear in the Bible. In medieval Europe, the northern lights were regarded with fear and superstition as omens of approaching disaster. Not surprisingly, the folklore of peoples of high northern latitudes, such as the Inuit, is rich in references to the aurora borealis. Similarly, the aurora australis occupies a prominent place in the mythology of the New Zealand Maoris as the "burning of the sky".

As with all optical phenomena, science has provided an explanation for auroras. We now know that they are caused when incoming electrons from solar emissions encounter gas molecules in the Earth's upper atmosphere, between 50 and 600 miles (80 to 1,000 km) above the surface. Traveling at roughly 1,000 miles (1,600 km) per second, these electrons collide with oxygen and nitrogen molecules in the atmosphere, producing an electronic "splash" of light, known as a quantum. The wavelength, and therefore the visible appearance of this light, depends on which molecule

The aurora australis viewed from the space shuttle Discovery.

The aurora borealis (above) and the aurora australis (right) appear in a variety of shapes and dazzling colors.

takes the impact of the electron, and the air pressure at which the collision occurs.

When electrons collide with oxygen molecules in low-pressure areas of the atmosphere, a yellow-green aurora occurs. Red is produced by collisions with oxygen in areas of even lower pressure at higher altitudes. A blue tinge is produced by interaction with atmospheric nitrogen.

Auroral Zones

Electrons have a negative charge and are directed north and south by the Earth's magnetic field toward the two magnetic poles. This is why auroras are primarily a high-latitude phenomenon, occurring within two "auroral zones", which lie about 20 to 25 degrees from the Earth's north and south magnetic poles. Seen from space, an aurora appears as a giant ring of glowing particles, centered on a magnetic pole.

Within the auroral zones, auroras are visible on most clear nights at any time of the year. They appear most frequently around the times of the equinoxes (see p. 24), for reasons that are not clear. What is better understood is the connection between auroras and sunspots (see p. 117). As auroras are caused by solar emissions, it is not surprising that they occur most often during the 11-year maximum in sunspot activity, and only rarely during the time of minimum sunspots. During periods of maximum activity, the northern lights have been seen as far south as Athens and Mexico City, and the southern lights as far north as Brisbane in Australia.

Attempts have been made over the years to confirm a link between intense solar activity and unusually cold or warm weather. Although tantalizing, such links have yet to be proved.

Green Flashes

On very rare occasions, when the Sun is on the point of disappearing below the horizon, a green light will be visible above it for a few seconds. This phenomenon is known as a green flash.

Sunlight consists of colors of different wavelengths, which are scattered by dust particles as they pass through the atmosphere. This causes the color of our sky to vary from blue through to red according to the amount of dust in the air and the path of the light through the atmosphere (see p. 56).

A similar effect causes the color of the Sun to change as it sets. Refraction of the Sun by the atmosphere creates a vertical spectrum of colors. These colors disappear below the horizon one by one, starting at the red end of the spectrum. For a brief moment after red, orange, and yellow have disappeared, green is the only color visible. The green light may be seen for just a short time, and in some cases, it may appear as a rapid flash.

WEATHER WATCH

Worldwide but best viewed over oceans at high latitudes

Just above the horizon

Refraction of direct sunlight by the atmosphere at sunrise or sunset

None

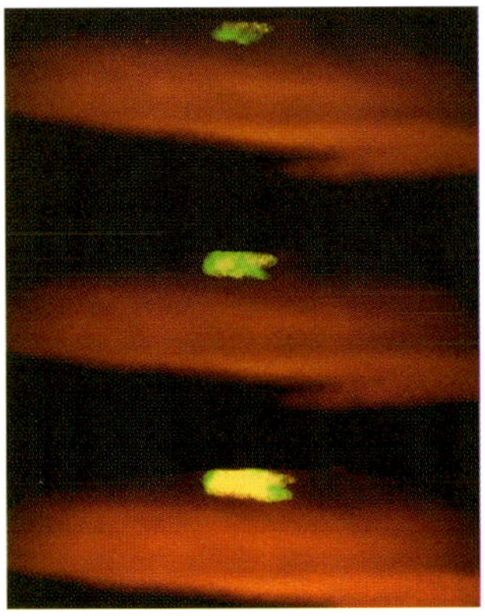

These time-lapse images (below) show the color of the Sun changing from yellow to green.

If the air were free of dust, there would be a blue and then a final violet flash. However, the atmosphere almost always contains sufficient dust particles to scatter blue and violet light. Moderate amounts of dust will also scatter green light, which is why this phenomenon is rare.

The sky around and above the Sun may remain red during this process, because light beyond the horizon continues to be scattered by particles in the air above.

This whole process may occur in reverse at sunrise. At sunrise or sunset, green flashes are best seen over the ocean, where the horizon is level and there is little dust, but they may also be viewed over flat land. They are more common in high latitudes, because there the sun rises and sets more slowly.

Since direct sunlight can cause lasting damage to eyesight, it is advisable to look at the setting or rising Sun for only very short periods.

Mirages

Mirages are produced when light is refracted as it passes through air layers of different temperatures and densities. The air acts like a lens, bending light and presenting a distorted, inverted, or enlarged image in a different position.

WEATHER WATCH

Usually over deserts or cold water

Close to the horizon

Refraction of light through air layers of different densities

Clear skies

There are two basic types of mirage—inferior and superior—although combinations of these can create a variety of effects.

The more common type is the inferior mirage. Normally, air density decreases with height, but when the surface of the ground is heated strongly, air density may actually increase for the first 10 to 20 feet (3 to 6 m). This often occurs in deserts on calm, clear summer days when surface temperatures rise sharply. Under these conditions, light moving downward will be refracted back upward, producing a false image (usually of an object on or near the horizon) just above the surface. Because the image actually forms nearer the viewer, it may appear to be situated below ground level. Thus, when an observer looks at a tree on the horizon, he or she will see both the true image and a shimmering, false image beneath it. Similarly, light from the sky may appear just below the horizon, creating an image like a lake.

A superior mirage, on the other hand, is produced when the air near the surface is colder (and denser) than the air immediately above it. Light heading upward will be refracted back down by the warm layer. This occurs most often over cold water and produces a false image above the level of (but nearer) the observer. A person looking at a ship on the water may therefore see an inverted image of it floating in the sky above.

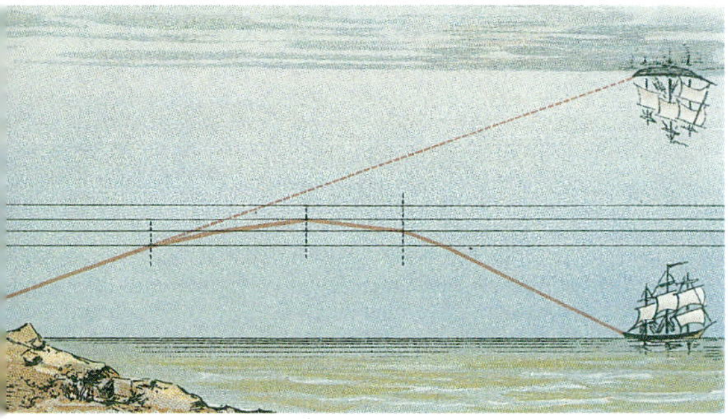

Inferior mirages occur often in arid areas such as the Namib Desert, Africa (above). This 1902 illustration (left) shows a superior mirage. While the viewer would see an inverted image like this, the upper image actually forms where the lines on the diagram converge.

Air Pollution

Air pollution is one atmospheric phenomenon that does not occur naturally: it is the direct result of human activity. Pollution comes from many sources, including industrial processes—particularly those that involve the burning of fossil fuels—and automobiles.

Pollution manifests itself in the atmosphere in the form of smog. Originally, this term was used to refer to a combination of smoke and fog, but now it is also used to refer to hazes that are caused by pollutants. In dry conditions, pollution will give rise to a haze that is basically a layer of smoke. However, the particles that make up the smoke dramatically increase the number of condensation nuclei in the air (see p. 41). This means that any water vapor present is more likely to condense and form smog.

WEATHER WATCH

Worldwide; common over large cities and industrial areas

Usually between the ground and 500–1,000' (150–300 m)

Emissions from factories and automobiles

Hazes and fogs, reduced visibility

Health hazard; may contribute to global warming

Smogs will clear only when a wind blows them away or rain disperses them. This means that the worst-affected cities are densely populated centers that receive little rainfall and regularly experience warm, clear, calm conditions, such as Mexico City and Los Angeles.

The exact type of haze or smog will vary according to the emissions. The most common constituents include carbon monoxide, nitrogen oxide, hydrocarbons, and sulfur dioxide. Many of these are harmful to human health. In particular, some of these substances can cause acidic smogs, which erode buildings and pose a serious threat to people with respiratory conditions. The burning of fossil fuels may also contribute to global warming (see p. 126).

Dense smogs caused by the burning of coal fires were a major problem in the United Kingdom from the late nineteenth century until the 1960s, when smokeless fuel was introduced.

Many industrial processes produce high levels of pollution.

Air pollution over Los Angeles (above) and the deadly London smog of 1952 (right).

One of the worst smogs occurred in London in 1952 and probably caused the deaths of around 4,000 people from respiratory diseases.

Trapped Air

The output of pollutants in urban areas is usually fairly uniform. However, certain meteorological conditions can greatly increase the impact of these pollutants.

The most significant of these conditions is the low-level temperature inversion. This occurs when a layer of warm air, often associated with a high-pressure system, moves over a layer of cold air and prevents it from rising and dispersing. Often, the result of this phenomenon is a dark band of smog extending from the ground to a height of around 500 to 1,000 feet (150 to 300 m).

As the day goes on, ground heating may erode the inversion layer from below, and once the inversion breaks, the smog may disperse fairly rapidly. However, if the inversion does not break, it is likely to intensify at night. This situation may persist for several days, raising pollution to critical levels.

A city's location and surrounding topographical features can also influence pollution patterns. In Los Angeles and Sydney, for example, smogs are blown inland by sea breezes and blocked by mountains behind the city.

Pollution levels also vary with seasonal weather patterns. On the west coast of the United States, during the summer months, inversions form when cool surface air from the ocean is overlaid by warmer air.

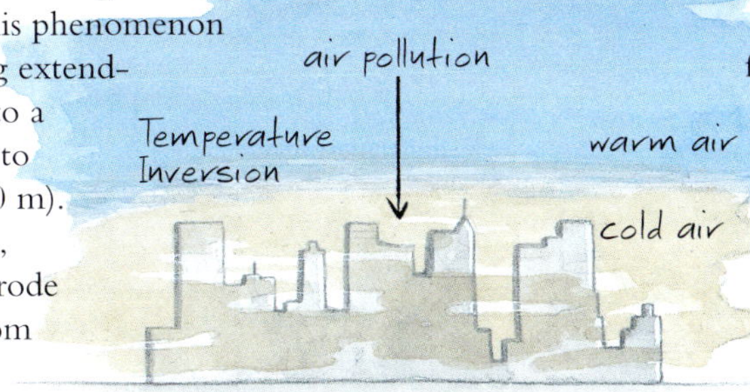

These meteorological factors mean that it is impossible to reduce pollution except through the control of emissions. In many cities restrictions on industrial output and automobile use are now being enforced.

Wildfires

Wildfires can have a dramatic effect on atmospheric visibility and their development is often related to weather patterns. The term wildfire is used to describe any uncontrolled fire burning through vegetation. Other expressions such as scrub fire and forest fire are used to describe fires in specific habitats. In Australia, all wildfires are known as bushfires.

WEATHER WATCH

Worldwide, common in California, southern France, Australia

Smoke may spread to upper troposphere

Burning of vegetation in favorable meteorological conditions

Can create pyrocumulus clouds that may produce rain and lightning

Danger to life and property

Wildfires occur most often in California, the French Riviera, and parts of Australia. These areas experience periodic drought, high summer temperatures and hot, dry winds, and are covered with highly volatile vegetation. These factors create ideal conditions for fires.

Although wildfires have always occurred naturally, usually as a result of lightning strikes, human activity has greatly increased their frequency, with out-of-control burn-off procedures, campfires, and discarded cigarettes being among the most common causes.

If a wildfire is not controlled at once, it can become an unstoppable inferno that rages across the countryside, causing loss of life and widespread property damage.

Fire in the Sky

Smoke from wildfires can rise high into the troposphere, increasing the number of condensation nuclei (see p. 41) in the air, and dramatically affecting sunlight. Normally, as light passes through the atmosphere, the colors of the spectrum are dispersed one by one, beginning at the violet end of the spectrum (see p. 56). During a fire, the invisible smoke particles

A firsthand view of a wildfire in France.

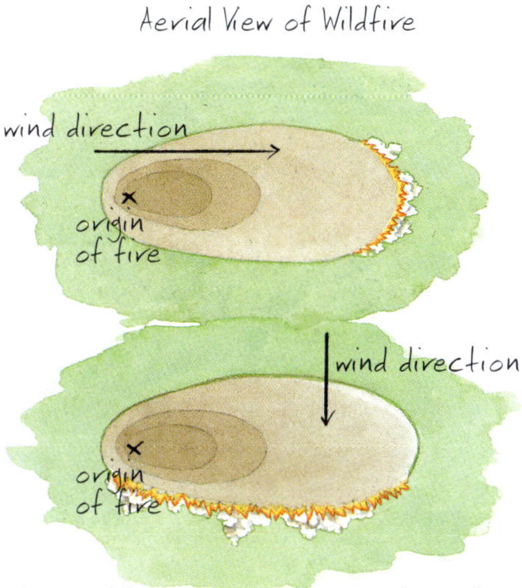

A cloud of smoke hangs over Sydney during widespread fires in 1994 (top). A satellite image of a fire in northern Australia (above).

enhance this scattering effect, so that colors toward the red end of the spectrum are scattered just above ground level, creating an eerie orange glow in the sky and intense red sunsets. This situation is likely to persist until the smoke is dispersed by wind or rain.

If a fire burns for some time, a pyrocumulus cloud (see p. 197) may form. This may produce lightning strikes that trigger further fires.

Once they have begun, wildfires may spread erratically. However, a knowledge of how fires interact with weather systems can help fire-fighters predict some of their movements.

If a fire begins and there is no wind, the fire will spread outward in all directions. If a wind then starts to blow across the fire, the fire will form an ellipse, with flames advancing slowly toward the wind, faster on the flanks, and even more rapidly downwind. This downwind fire-front is the most dangerous part of the fire. Any sudden change in wind direction, perhaps associated with the arrival of a frontal system, can create a wider and more threatening front along the side of the ellipse. For this reason, firefighters will seek regular advice from the local weather bureau.

To anyone watching the fire, the direction in which the smoke is blowing will indicate the location of the fast-moving firefront. A significant increase in ground-level smoke near the observer is a warning that it is time to leave the area.

Volcanic Clouds

During prehistoric times, volcanic activity was far more frequent and violent than it is now, and the gases belched into the sky from volcanoes—including carbon dioxide and oxygen—contributed to the formation of our atmosphere. Volcanic clouds had a significant effect on the climate, reducing temperatures and intensifying precipitation. Nowadays, major eruptions occur only from time to time, but the power of these eruptions is so immense that they still have a short-term influence on the weather, and their effects are seen worldwide.

WEATHER WATCH

Effects experienced worldwide

Volcanic clouds can extend from surface to 60,000' (18,000 m)

Volcanic eruptions forcing ash and dust into atmosphere

Volcanic clouds may produce acid rain; eruption often accompanied by lightning

Danger to life and property in immediate vicinity; aircraft hazard

Volcanic eruptions eject huge amounts of ash and dust particles into the atmosphere and these are dispersed around the globe. In a powerful eruption, some of this material may pass through the tropopause into the stratosphere. Stratospheric winds may then keep the ash and dust circulating around the Earth for several years.

Large ash clouds have a significant effect on weather patterns because they reflect a certain amount of incoming solar radiation back into space, inhibiting the heating effect of the Sun. This was first noticed by Benjamin Franklin after a volcanic eruption in Iceland in 1783, when he discovered that sunlight directed through a magnifying glass would no longer set fire to a piece of paper.

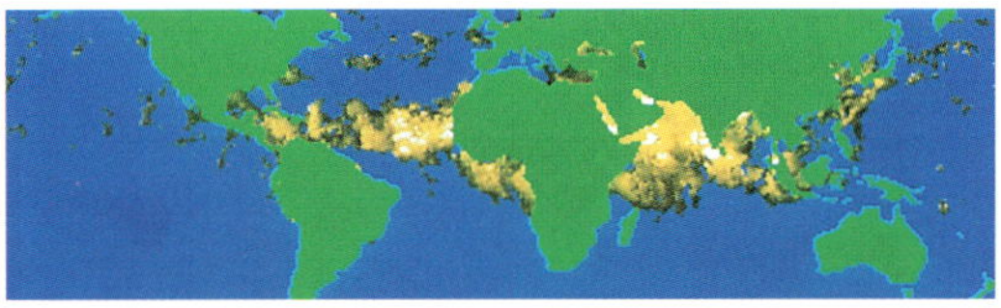

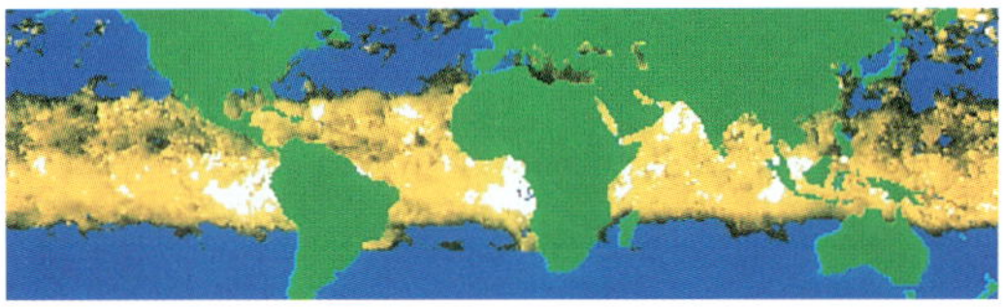

These satellite images show, in yellow, the distribution of volcanic material in the atmosphere immediately after the eruption of Mount Pinatubo in the Philippines in 1991 (above left) and two months later (left). The ash gradually spread around the globe, forming a wide band over low latitudes and creating dramatic sunsets and sunrises (top) in many parts of the world.

In Europe, the year after the greatest eruption on record, on the island of Tambora near Borneo in 1815, became known as the "year without a summer". Similarly, the eruption of Mount Pinatubo in the Philippines in 1991 created a sulfuric acid haze in the stratosphere that later spread around the globe. This probably had a cooling effect on world climate that may have lasted for more than two years. However, while temporary cooling almost certainly occurs because of volcanic eruptions, it is difficult to separate this effect from the fluctuations in temperature that occur as part of our normal weather patterns. Nor is it clear whether any other variables, such as rainfall, are affected.

Ash and smoke from Mount St Helens darkened skies across the United States for days after the eruption on 18 May 1980.

Volcanic Twilights

Volcanic dust has a significant effect on our blue skies and red sunsets. When sunlight passes through our atmosphere, particles in the air scatter the colors of the spectrum. The colors we see change according to the angle at which the light passes through the atmosphere, which varies with the time of day (see p. 56). The addition of volcanic dust to the atmosphere increases the number of particles and hence the scattering of colors. This means that colors at the red end of the spectrum are more effectively dispersed across the sky, producing paler daytime skies and intense red and purple sunsets (known as volcanic twilights) and sunrises.

Volcanic eruptions can be a major hazard to aviation, because the small particles thrust into the atmosphere may clog airplane engines. Several recent near-disasters led to the founding of an international aviation warning system, run by the World Meteorological Organization (WMO). Satellite images of volcanic clouds are studied together with forecast dispersal patterns, so that no-go areas for aircraft can be established.

Sunshine is delicious, rain is refreshing, wind braces us up, snow is exhilarating; there's really no such thing as bad weather, only different kinds of good weather.

JOHN RUSKIN (1819–1900),
English writer and critic

Resources Directory

FURTHER READING

General

Atmosphere, Weather and Climate, by Roger G. Barry and Richard J. Chorley (Routledge, 1968). An excellent textbook for advanced readers.

The Audubon Society Field Guides: Weather, by David M. Ludlum (Alfred A. Knopf, 1993). A conveniently sized, comprehensive field guide to weather phenomena.

The Australian Weather Book, by K. Colls and R. Whitaker (National Book Distributors, 1993). A detailed guide to Australia's climate and weather.

Climate, History and the Modern World, by H. H. Lamb (Routledge, 1995). An examination of the connections between climate change and historical developments.

Essentials of Meteorology, by C. Donald Ahrens (West Publishing Company, 1993). A beautifully illustrated introduction to meteorology, with a US focus.

Fundamentals of Meteorology, by Louis J. Battan (Prentice Hall, 1984). An introduction to meteorology, with a US focus.

General Climatology, by H. J. Critchfield (Prentice Hall, 1983). This textbook covers most aspects of meteorology and reviews their social and economic implications.

Meteorology Today, by C. Donald Ahrens (West Publishing Company, 1985). An excellent, well-illustrated introduction to weather, with a US focus.

The Science and Wonders of the Atmosphere, by Stanley David Gedzelman (Wiley, 1980). A comprehensive textbook, incorporating interesting historical material.

Spacious Skies—The Ultimate Cloud Book, by Richard Scorer and Arjen Verkaik (David & Charles, 1989). An excellent description of cloud types and the processes leading to their formation.

The USA Today Weather Almanac 1995, by Jack Williams (Vintage Books, 1994). A detailed look at the USA's weather, including weather profiles of 200 cities and a range of weather records.

Watching the World's Weather, by William J. Burroughs (Cambridge University Press, 1991). Analyzes how weather satellites have transformed the science of meteorology. Includes information on how to interpret satellite images.

Weather, by Sally Morgan (Time-Life, 1996). Part of the Nature Company Discoveries series for children, this is a superbly illustrated and easy-to-understand introduction to the weather.

Weather, by Brian Cosgrove (HarperCollins, 1992). An introduction to the weather for the young reader.

The Weather Book, by Ralph Hardy, Peter Wright, John Gribbin, and John Kington (Michael Joseph, 1982). An illustrated guide to meteorological phenomena and climates. Includes historical material.

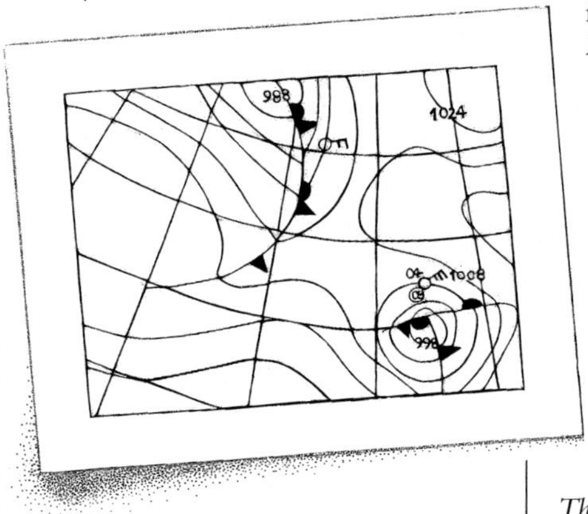

The Weather Book—An Easy-to-Understand Guide to the USA's Weather, by Jack Williams (Vintage Books, 1992). A first-rate introduction to basic meteorological concepts and the weather of the USA.

Weather Cycles: Real or Imaginary, by William J. Burroughs (Cambridge University Press, 1994). Explores the unresolved debate on the existence of weather cycles and discusses its implications for understanding climate change.

The Wonders of the Weather, by Bob Crowder/Bureau of Meteorology (Australian Government Publishing Service, 1995). An easy-to-read introduction to meteorology, with a strong Australian focus. Beautifully illustrated.

History of Meteorology

A Century of Weather Service, by Patrick Hughes (Gordon and Breach, 1970). An excellent review of the history of the US National Weather Service from 1870 to 1970.

Forty Years of Progress and Achievement—A Historical Review of WMO, edited by Sir Arthur Davies (World Meteorological Organization, 1990). Reviews the history of the IMO and WMO.

Manual of Meteorology: Vol. 1—Meteorology in History, by Sir Napier Shaw with the assistance of Elaine Austin (Cambridge University Press, 1942). A scholarly review of the history of meteorology from ancient times to 1920.

Meteorology—A Historical Survey (Vol. 1), by A. Kh. Khrgian, translated from Russian and edited by Ron Hardin (US Department of Commerce and The National Science Foundation, 1970). A meticulously researched history of meteorology from ancient times to about 1920, with a brief overview of the period between 1920 and 1958.

The Weather Bureau—Its History, Activities and Organization, by Gustavus A. Webber (D. Appleton and Co., 1922). A history of the US National Weather Service.

Weather Folklore and Local Weather Signs, by Edward B. Garriott (US Weather Bureau, US Dept of Agriculture, 1908). Includes local weather lore for 143 US cities.

Weather Lore—A collection of proverbs, sayings and rules concerning the weather, by Richard Inwards (E. Stock, 1898). Includes many proverbs and sayings concerning weather phenomena, and has an extensive bibliography. There is also an earlier, shorter edition, entitled *Weather Wisdom*.

Climates and Biomes

The Amateur Naturalist, by Gerald and Lee Durrell (Alfred A. Knopf, 1992). A naturalist's guide to 17 different environments.

The Audubon Society Nature Guides (Alfred A. Knopf). A series of well-illustrated guides to various North American habitats.

Desert Wildlife, by Edmund Jaeger (Stanford University Press, 1961). A series of well-written essays on North American desert animals.

Diversity and the Tropical Rain Forest, by John Terborgh (Scientific American Library, 1992). Several essays cover the ecological and evolutionary basis for the diversity of rain-forest organisms.

Ecology and Field Ecology, by Robert Leo Smith (Harper & Row, 1980). A college text with excellent coverage of the world's biomes.

The Fight for Survival: Animals in their Natural Habitats, edited by Peter Brazaitis and Myrna Watanabe (Michael Friedman Publishing Group, 1994). A well-illustrated book on how animals adapt to different habitats.

Grizmek's Animal Life Encyclopedia, edited by Bernhard Grizmek (13 vols) (Van Nostrand Reinhold Co., 1975). Comprehensive coverage of the world's animal species.

Habitats, edited by Tony Hare (Duncan Baird Publishers, 1994). A large-format book with 14 foldout panoramas of the world's ecological zones.

Land Above the Trees: A Guide to the American Tundra, by Ann Zwinger and Beatrice Willard (Harper & Row, 1972). Describes the ecology of the North American tundra, with an emphasis on plants.

The Living Planet, by David Attenborough (Reader's Digest, 1985). A superbly illustrated, habitat-based look at the world's wildlife. Based on the successful TV series.

Mountains: A Natural History and Hiking Guide, by Margaret Fuller (John Wiley & Sons, 1989). Discusses the geology, weather, wildlife, and ecology of the world's mountains.

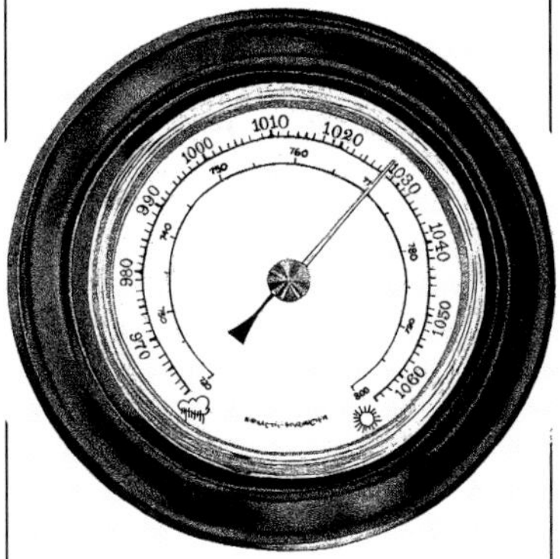

Natural History of the Antarctic Peninsula, by Sanford Moss (Columbia University Press, 1988). Beautifully illustrated with original drawings.

Portraits of the Rainforest, by Adrian Forsyth (Camden House, 1990). A series of essays on ecological aspects of rain-forest organisms.

The Sierra Club Naturalist's Guides (Sierra Club Books). A series of conveniently sized field guides covering the natural history of many North American habitats.

Wildlife of the Mountains, by Edward R. Ricciuti (Harry N. Abrams, 1979). Deals with the adaptations of mountain animals. Excellent photographs.

Magazines

For the addresses of the organizations mentioned below, see pp. 276–7.

The Australian Meteorological Magazine.
The journal of the Bureau of Meteorology, Australia. Published quarterly.

Bulletin of the American Meteorological Society.
A monthly journal covering the activities of the American Meteorological Society.

International Journal of Climatology.
A monthly journal of the Royal Meteorological Society.

Meteorological Applications.
A quarterly journal of the Royal Meteorological Society.

National Weather Digest.
A quarterly journal of the National Weather Association (NWA), dealing mainly with weather forecasting.

Weather.
A monthly journal of the Royal Meteorological Society.

Weatherwise.
A popular, non-technical, bimonthly journal dealing with meteorological issues in the US.
Available through:
Heldref Publications,
1319 Eighteenth Street NW,
Washington, DC 20036, USA.

WMO Bulletin.
A quarterly journal on the activities of the WMO.
Available through:
Unipub,
Box 433,
Murray Hill Station,
New York, NY 10016, USA.

Computer Services

The worldwide computer network, the Internet, contains a number of sites specializing in weather-related topics. Several useful sites are listed below.

Atmoslist: A distribution list for Australian atmospheric scientists and those working in closely related fields.
http://www.monash.edu.au/atmos/

NASA: Satellite images of the Earth and its atmosphere.
http://www.hq.nasa.gov/

National Climatic Data Center (NCDC): Climatic data for US and the world.
http://www.ncdc.noaa.gov/

National Oceanic and Atmospheric Administration (NOAA): Forecasts, satellite images, and weather maps.
http://www.noaa.gov/

The Weather Channel Forum: News, statistics, and forums. Available only on Compuserve.
GO.TWCFORUM

WeatherNet: A list of 250 Internet weather sites, updated regularly.
http://cirrus.sprl.umich

The Weather Page: A comprehensive list of weather services on the Internet.
http://acro.harvard.edu/GA/weather.html

ORGANIZATIONS

International

World Meteorological Organization (WMO), 41 Avenue Giuseppe-Motta, Geneva, Switzerland; tel. (41 22) 730 8315 fax (41 22) 733 0242 e-mail: ipa@www.wmo.ch
Most countries are members. Coordinates World Weather Watch, the World Climate Program, and provides many services, including assistance to developing countries. Publishes a quarterly bulletin.

International Weather Watchers, PO Box 77442, Washington, DC 20013, USA tel. (202) 544 4999 e-mail: iww@delphi.com
Society for both amateur and professional enthusiasts. Publishes bimonthly magazine and organizes meetings.

Australia

Bureau of Meteorology, GPO Box 1289K, Melbourne 3001; tel. (613) 9669 4000 fax (613) 9669 4113 e-mail: gopher://babel.ho.bom.gov.au/1
Head office in Melbourne (open to public) with regional forecasting centers based in other Australian cities. Publishes a range of material.

The Australian Meteorological and Oceanographic Society (AMOS), PO Box 654E, Melbourne 3001; tel. (613) 9586 7658 fax (613) 9586 7600 e-mail: vxj@dar.csiro.au
Membership mostly professional, but open to amateurs. Publishes a bimonthly bulletin. Offices in Hobart, Melbourne, and Sydney.

Canada

Atmospheric Environment Service, Environment Canada, Inquiry Center, Ottawa, Ontario K1A 0H3; tel. (800) 668 6767 Envirofax (819) 953 0966 e-mail: http://www.doe.ca
National offices in Toronto, Dorval, and Hull. Provides information and carries out research. Fact sheets available on topics such as tornadoes, thunderstorms, climate change, and the ozone layer.

The Canadian Meteorological and Oceanographic Society (CMOS), 903/151 Slater St, Ottawa, Ontario, K1P 5H3; tel. (613) 990 0300 fax (613) 990 5510 e-mail: CMOS@ottmed.meds.dfo.ca
13 centers across Canada hold formal and informal meetings on meteorological and oceanographic subjects. The society publishes a number of journals.

France

Météo-France, 1 quai Branly, 75340 Paris Cedex 07; tel. (331) 4556 7425 fax (331) 4556 7111 e-mail: http://www.meteo.fr/
Operates regional and local services for French territory, both in metropolitan France and overseas. A catalog of its many publications is available.

Société météorologique de France (SMF), 1 quai Branly, 75340 Paris Cedex 07; tel. (331) 4556 7364 fax (331) 4556 7363
Scientific society of both professionals and amateurs. Publishes a quarterly journal.

Germany

Deutsche Meteorologische Gesellschaft (DMG), Mont Royal, D-56841 Traben-Trarbach; tel. (49) 654 1180.
Members include professionals and amateurs. Publishes a magazine.

Deutscher Wetterdienst, Stabsstelle Offentlichkeitsarbeit/Pressesprecher, Postfach 10 04 65, D-63004 Offenbach; tel. (49 69) 8062 2294 fax (49 69) 8962 2488
Business-related and media forecasting services.

Meteo Consult, Konrad-Adenauer-Strasse 30A, D-55218 Ingelheim/Rhein; tel. (499) 613278060
Private weather and climate consultation services.

More + More Consult, Hohenadelstrasse 2, D-85737 Ismaning; tel. (49) 899613927
Private company issuing weather advice.

Hong Kong

Royal Observatory, Hong Kong, 134A Nathan Rd, Kowloon; tel. (852) 2926 8200 fax (852) 2721 5034 e-mail: metgeos@hk.net
Forecasts weather, including tropical cyclones, and monitors radiation levels in the environment. Publishes and holds a large number of scientific papers.

India

India Meteorological Department, Lodi Road, New Delhi 110003; tel. (91 11) 461-9415 fax (91 11) 469-9216
Meteorological and climatic information available. The department's cyclone warning service is one of the oldest in the world.

India Meteorological Department (Research), Shivaji Nagar, Pune 411005; tel. (91 212) 325797 fax (91 212) 323201
Weather forecasting and research center. Holds meteorological records for the past 100 years.

Japan

Japan Meteorological Agency (JMA), 1-3-4, Otemachi, Chiyoda-ku, Tokyo 100; tel. (813) 3211 4966 fax (813) 3211 2032
Issues short- and long-range forecasts, including warnings of typhoon, tsunami, storm surge, flood, earthquake, and volcanic activity.

New Zealand

Meteorological Service of New Zealand Limited (MetService), PO Box 722, Wellington 6015; tel. (644) 472 9379 fax (644) 499 1942 e-mail: service@met.co.nz Weather information and forecasting services to media, industry, aviation, agriculture.

Meteorological Society of New Zealand, PO Box 6523, Wellington; tel. (644) 386 0300 fax (644) 386 2153 Members include professional and amateur enthusiasts. Publishes a quarterly newsletter and biannual journal.

National Institute of Water and Atmospheric Research (NIWA), PO Box 14901, Kilbirnie, Wellington; tel. (644) 386 0300 fax (644) 386 2153 Investigates weather, climate, and atmosphere, and publishes a quarterly bulletin.

Singapore

Singapore Meteorological Service, PO Box 8, Changi Airport, Singapore 9181; tel. (65) 545-7190 fax (65) 545-7192 e-mail: metsin@cs.gov.sg Provides weather and climatic information for government, business, and general public.

South Africa

South African Weather Bureau, Information and Publication Section, Private Bag X097, Pretoria 0001; tel. (27 12) 290 8025 fax (27 12) 290 2958 e-mail: climenq@cirrus.sawb.gov.za Operates 16 regional forecasting services, including Marion and Gough Islands in the southern oceans.

United Kingdom

Royal Meteorological Society, 104 Oxford Rd, Reading, Berkshire RG1 7LJ; tel. (441 734) 568500 fax (441 734) 568571 Publishes several journals for weather enthusiasts, organizes conferences, and is also involved with educational work. Members include both working scientists and weather enthusiasts in other occupations.

The Met. Office, Bracknell, Berkshire RG12 1AA; tel. (44 1344) 854455 fax (44 1344) 854942 The UK national weather service. As well as developing computer-based weather forecasting, this office provides information on such vital issues as global climate change and ozone depletion.

European Centre for Medium-Range Weather Forecasts (ECMWF), Shinfield Park, Reading, Berkshire RG2 9AX; tel. (44 1734) 499 104 fax (44 1734) 869 450 e-mail: dra@ecmwf.int Provides medium-range forecasts (1 to 10 days) for its 17 member countries and other services around the world.

United States

American Meteorological Society (AMS), 45 Beacon St, Boston, Massachusetts 02108-3693; tel. (617) 227 2425 fax (617) 742 8718 e-mail: amsinfo@ametsoc.org More than 11,000 members, mostly professional meteorologists, oceanographers, and hydrologists. Publishes seven journals, a monthly bulletin, and occasional monographs.

Commercial Weather Services Association, 655 Fifteenth St, NW, Suite 310, Washington, DC 20005-5701; tel. (202) 546 6993 fax (202) 546 2121 Provides tailored weather, river, and water resources forecasts and hydrometeorological consultation.

For Spacious Skies, 54 Webb Street, Lexington, Massachusetts 02173; tel. (617) 862 4289 A non-profit organization dedicated to fostering awareness of the wonders of the sky.

National Climatic Data Center, 151 Patten Ave, Room 120, Asheville, North Carolina 28801-5001; tel. (704) 271 4800 fax (704) 271 4876 Provides climatic data to industry and the general public.

National Oceanic and Atmospheric Administration (NOAA), National Weather Service, 1305 East–West Highway, 8624 Floor, Silver Spring, Maryland 20910; tel. (301) 713 1208 e-mail: http://www/nws.noaa.gov/ Operates forecasting services nationally.

National Weather Association (NWA), 6704 Wolke Court, Montgomery, Alabama 36116-2134; tel./fax (334) 213 0388 e-mail: natweaasoc@aol.com Professional association promoting excellence in meteorology and related activities to an international membership. Publishes a monthly newsletter and a quarterly journal. Anyone with an interest in weather is welcome to join.

INDEX *and* GLOSSARY

In this combined index and glossary, bold page numbers indicate the main reference, and italics indicate illustrations and photographs.

C

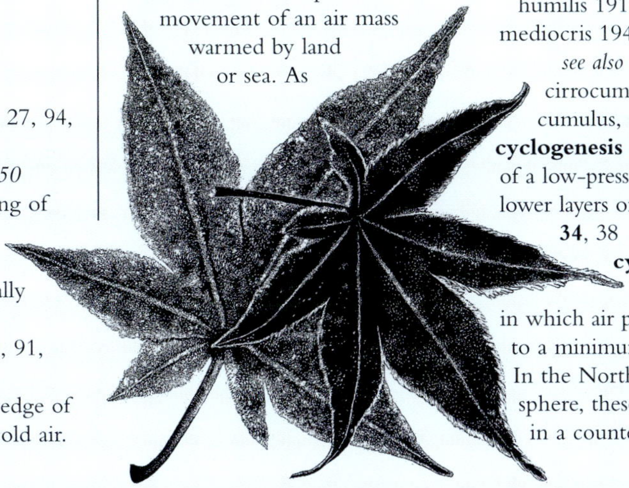

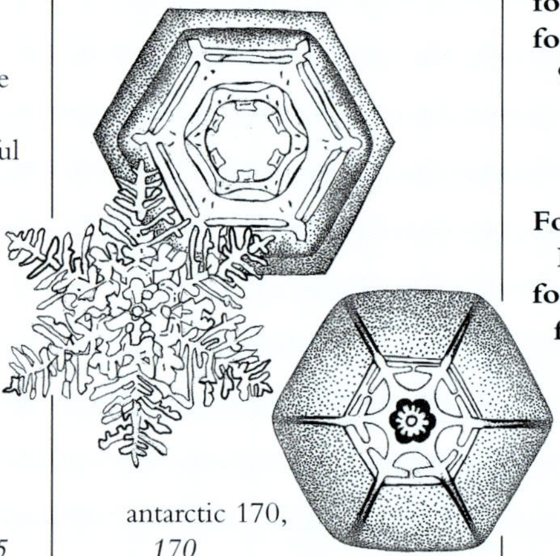

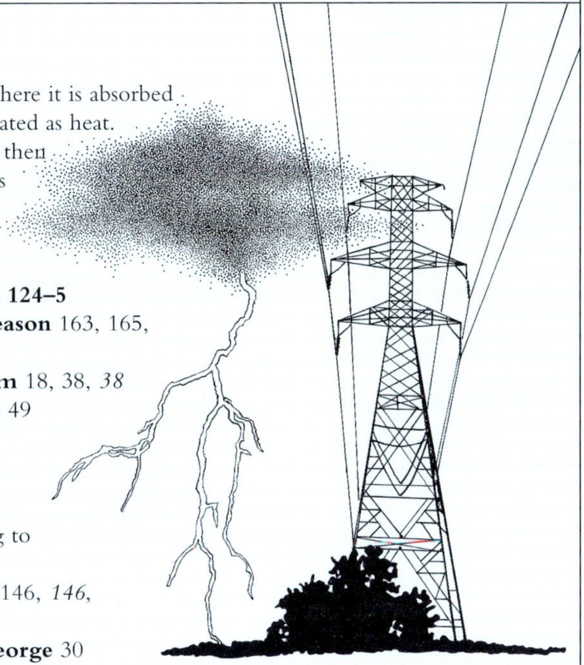

H

I

N
E
W
S

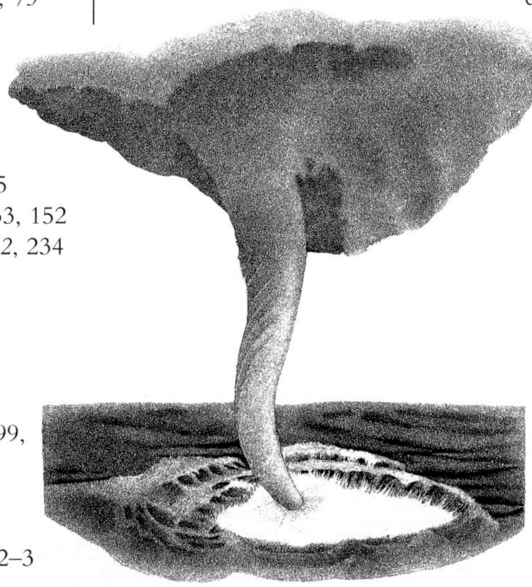

CONTRIBUTORS

William J. Burroughs lives in the United Kingdom and has written nearly 200 articles and papers on all aspects of the weather and climate change. His books include *Watching the World's Weather* and *Weather Cycles: Real or Imaginary?* He has a PhD in Atmospheric Physics and has held a series of senior posts in the Department of Energy and the Department of Health.

Bob Crowder joined the Australian Bureau of Meteorology in 1951 and went on to hold a number of positions there until retiring as the Deputy Director (Services) in 1988. He was one of Australia's first TV weather presenters, appearing regularly from 1958 to 1962. He has extensive experience in international meteorology, including work with the World Meteorological Organization (WMO) Commission for Basic Systems. His book, *The Wonders of the Weather*, was published in 1995.

Ted Robertson is a biologist and teacher at the University of California's Lawrence Hall of Science. In his spare time, he writes science books, teaches wilderness survival in the mountains and deserts of California, and assists conservation organizations with environmental impact reports. He has led natural history trips to many of the world's biomes, including the arctic and alpine tundra; tropical, temperate, and montane forests; and the deserts, grasslands, and coastal regions of North and South America.

Eleanor Vallier-Talbot is a meteorologist and Outreach Coordinator with the National Weather Service in Boston, Massachusetts, USA. She is actively involved in a number of groups encouraging amateur weather enthusiasts, including the International Weather Watchers (IWW).

Richard Whitaker completed an honors degree in science at Monash University, Melbourne, Australia in 1968. He then moved to Sydney and joined the Bureau of Meteorology. He is now the New South Wales manager of the Special Services Unit, where he is responsible for the bureau's commercial activities in that state.

CAPTIONS

Page 1: Altostratus cloud over woodlands at sunset, UK.
Page 2: Beech trees in a winter mist, UK.
Page 3: Frosted feather on red oak leaves, UK.
Pages 4–5: Lightning over Monument Valley, Arizona, USA.
Pages 6–7: A lenticular cloud over the Three Sisters in Central Oregon, USA, reflected in melting ice.
Pages 8–9: Storm clouds gather over Amboseli National Park, Kenya, Africa.
Pages 10–11: Colors streak the sky over the Garden State Parkway, New Jersey, USA.
Pages 12–13: The Golden Gate Bridge enveloped by fog over San Francisco Bay, USA.
Pages 22–3: A tornado emerges from the base of a cumulonimbus cloud over midwestern USA.
Pages 60–1: A hot-air balloon rises toward strato-cumulus clouds at sunset.
Pages 78–9: From space, the eye of this massive hurricane is clearly visible.
Pages 106–7: A frost fair, held on the frozen Thames River, during the Great Frost of 1739–40.
Pages 128–9: Two camel riders in the Thar Desert, Rajasthan, India.
Pages 142–3: A group of emperor penguins stands out against the stark, white antarctic landscape.
Pages 174–5: A waterspout and a bolt of cloud-to-ground lightning appear simultaneously in an awesome natural display.
Pages 178–9: A veil of morning fog cloaks countryside in Western Australia.
Page 179 (inset top): Strawberry leaves edged with frost, Michigan, USA.
Page 179 (inset bottom): Morning frost covers uplands in Wales, UK.
Pages 188–9: A majestic cumulonimbus cloud seen at sunset over the ocean.
Page 189 (inset top): Altocumulus undulatus clouds are particularly beautiful at sunset.
Page 189 (inset bottom): A stratocumulus cloud against a bright blue sky.
Pages 218–9: Car headlights in heavy rain.
Page 219 (inset top): Rainfall over Myall Lakes, New South Wales, Australia.
Page 219 (inset bottom): A fence in Central Oregon, USA, covered in snow and ice after a heavy storm.
Pages 236–7: Multiple bolts of cloud-to-ground lightning during a storm over southwestern USA.
Page 237 (inset top): Palm trees blown by a hurricane over Plantation Island, Fiji.
Page 237 (inset bottom): A tornado in midwestern USA.
Pages 254–5: A rainbow over the Banda Sea, Indonesia.
Page 255 (inset top): Aurora borealis, seen from Wisconsin, USA.
Page 255 (inset bottom): Iridescence in altostratus cloud.
Pages 272–3: This etching from 1846 depicts various atmospheric and meteorological phenomena.

ACKNOWLEDGMENTS

The publishers wish to thank the following people for their assistance in the production of this book: Ionas Kaltenbach, Margaret McPhee, Di Regtop, Oliver Strewe, and Luke Tyler and the pupils of St Ignatius' College, Riverview, Sydney, Australia.

PICTURE AND ILLUSTRATION CREDITS

(t = top, b = bottom, l = left, r = right, c = centre, i = inset
A = Auscape International; AA/ES = Animals Animals/Earth Scenes; AA&A = Ancient Art and Architecture Collection, London; APL = Australian Picture Library; Backgrounds = Backgrounds Photo Library; BCL = Bruce Coleman Limited, UK; Bettman = Bettman Archive; Bridgeman = Bridgeman Art Library, London; FLPA = Frank Lane Picture Agency; Granger = The Granger Collection, New York; IS = International Stock Photo; LT = Lochman Transparencies; ME = Mary Evans Picture Library; MP = Minden Pictures; NCAR = National Centre for Atmospheric Research/University Corporation for Atmospheric Research/National Science Foundation; NHPA = Natural History Photographic Agency; OSF = Oxford Scientific Films; PE = Planet Earth Pictures; PL = The Photo Library -Sydney; PR = Photo Researchers; SAL = Survival Anglia; SPL = Science Photo Library; TS = Tom Stack and Associates; Werner = Werner Forman Archive; W = Wildlight Photo Ag.)

1 WS Pike/SAL/OSF **2** EA Janes/NHPA **3** William Paton/SAL/OSF **4-5** Chad Ehlers/IS **6-7** Bob Pool/TS **8-9** Martyn Colbeck/OSF **10-11** Herb Segars/AA/ES **12-13** Hilary Wilkes/IS **14**tl Schindler Collection, New York/Werner; tr Paul McCullagh/OSF; cr Bob Firth /IS **15**t John Downer/PE; c John Eastcott and Yva Momatiuk/PE; b John Shaw/A **16**t D Hoadley/FLPA; bl Warren Faidley/IS; br SPL/PL **17**t Mark Marten/NASA/PR; c Bob Firth /IS; b John Downer/OSF **18** Joel Bennett/SAL/OSF **19** ER Degginger/AA/ES **20**t Bettman/APL; c Galen Rowell/Hedgehog House, New Zealand; b J Robert Stottlemyer/IS **21**t Hank Morgan/SPL/PL; c Larry Lipsky/TS; b SPL/PL **22-23** Warren Faidley/IS **24**tl ME; b Mark Marten/NASA/PR **26**t Chris Curry/Hedgehog House, New Zealand; c Thomas Kitchin/TS; cr Franca Principe/Istituto e Museo di Storia della Scienza di Firenze **27**l David E Rowley/PE; r Ian Murphy/Tony Stone Worldwide/PL **28**t Biblioteca Estense, Modena/Bridgeman; bl World Perspectives/Tony Stone Images/PL; bc John Eastcott and Yva Momatiuk/PE **29** Granger **30**t Jean-Loup Charmet; c Bill Rossow/NASA **31**t NASA/SPL/PL; b Jean-Loup Charmet **32**t Austin J Brown/Aviation Picture Library; c Stan Osolinski/OSF **33**t NASA/SPL/PL; b JHC Wilson/Robert Harding Picture Library **34**l David Miller; r WS Pike/SAL/OSF **35**c ESA/SPL/PL **36**t Peter Jarver/Backgrounds; c Stan Osolinski/OSF; b William M Smithey Jr/PE **37** George Ranalli/PR **38**b Los Alamos Nat Lab/SPL/PL **39**t R Sorensen and J Olsen/NHPA **40**tl John Shaw/ NHPA **40-41** Eric Soder/NHPA **41**t Flip de Nooyer/BCL; cr Rod Planck/TS; b Franca Principe/Istituto e Museo di Storia della Scienza di Firenze **42**c S McCutheon/FLPA **43**t Peter Jarver/Backgrounds; c Keith Gunnar/BCL **44**cr Dennis Sarson/LT; bl Kenneth E Woodley/Royal Meteorological Society/National Meteorological Library **45**tl Stephen Krasemann/NHPA; c Dr ER Degginger **46**t EPI Nancy Adams/TS; c Rod Planck/TS; b ME **47** Peter Davey/BCL **48** Daniel J Cox/Natural Exposures **49**t Peter Jarver/Backgrounds; c B Cosgrove/FLPA; b SPL/PL **50** ER Degginger/PL **51**t Chad Ehlers/IS; cr NASA/SPL/PL; b Bettman/APL **52**t Dennis Fisher/IS; b H Hoflinger/FLPA **53** Sheila Beougher/W/Liaison **54** Jeff Greenberg/IS **55**t NOAA/SPL/PL; c Warren Faidley/IS; b ME **56** Dries van Zyl/BCL **57** Col Roberts/LT **58**t Ashmolean Museum, Oxford; c David M Dennis/TS; b Tom Till/A **59**t Gerald Cubitt/BCL; cl GD Plage/BCL; cr NCDC/NOAA; b Dick Smith/Hedgehog House, New Zealand **60-61** John Shaw/NHPA **62**tl Thjodminjasafn, Reykjavik, Iceland (National Museum)/Werner; r AA&A; b National Museum of Anthropology, Mexico City/Werner **63**tl AA&A; c Museum of Anthropology, University of British Columbia, Vancouver/Werner; b Scala **64**tl Granger; c ME; b Palazzo del Té Mantova/Scala **65**tl and b Jean-Loup Charmet; tr British Library **66**tl and r Franca Principe/Istituto e Museo di Storia della Scienza di Firenze; b North Wind **67**t Franca Principe/Istituto e Museo di Storia della Scienza di Firenze; c Granger; b (by Martellini Gaspero) Tribuna di Galileo Firenze/Scala **68**tl Bettman/APL; tr ME; b Jean-Loup Charmet **69**tc Jean-Loup Charmet; tr and b Granger **70**tr Franca Principe/Istituto e Museo di Storia della Scienza di Firenze; c Granger; bl David Miller; bc Jiri Lochman/LT **71**t Granger; b (by Sano di Pietro) Biblioteca Comunale Siena/Scala **72**tr Christie's, London/Bridgeman; cl Royal Naval College, Greenwich/Bridgeman; cr Granger; b ME **73**tl World Meteorological Organization; tr Research Library/Australian Museum; b Jean-Loup Charmet **74**tl Granger; c SPL/PL; b ME **75**tl

Clive Collins; tr United States Air Force Geophysics Directorate; c Hulton Deutsch/PL; b George W Platzman/University of Chicago **76**tl World Meteorological Organization; tr Lafayette/National Meteorological Library; cl and b Hulton Deutsch/PL **77**t The Science Museum/Science and Society Picture Library; b UPI/Bettman/APL **78-79** NASA/TSADO/TS **80**tl Royal Meteorological Society; tr Australian Bureau of Meteorology; c Paul Nevin/PL; b Japan Meteorological Agency **81**tl Mark Newman/FLPA; tr Paolo Koch/PR; cr NOAA; br Mark Burnett/PR **82**tl IS; tr Paul Nevin/PL; cr Reg Morrison; b D Parer and E Parer-Cook/A **83**t NCAR; c SPL/PL; bl Thomas Bettge/NCAR **84** National Weather Service/NOAA **85** Paul Nevin **86**tl Oliver Strewe; bl Sony Australia Ltd and The Weather Channel bc NBC News **86-87** Rainbird/Robert Harding Picture Library **87**t JM LaRoque/A; bl and br Warren Faidley/IS **88**tl Elliott Varner Smith/IS; tr Tom McHugh/PR; bl GR Roberts/Documentary Photographs, br Oliver Strewe **89**t and c NCDC/NOAA; b Paula Bronstein/The Hartford Courant **90**t Clay Bryce/LT; bl WS Pike/SAL/OSF; br Peter Jarver/Backgrounds **91**t John Shaw/NHPA; cl Bob McKeever/TS; cr Maxwell Optical Industries P/L; b David Miller **92**tl Gary Milburn/TS; tr Rod Planck/TS; cr Peter Dean/FLPA; b Leonard Lee Rue III/PR **93**c Michael Marten/SPL/PL; bl Joe McDonald/TS; br Flip de Nooyer/BCL **94**t and b Oliver Strewe **95**tl Kevin Rushby/BCL; tr Taylor/Fairfax Photo Library, Sydney; b Solar Flair P/L/Davis Instruments **96**tl G Tomsich/Istituto e Museo di Storia della Scienza di Firenze/PR; c and b Oliver Strewe **97**t Solar Flair P/L/Davis Instruments; c and b Oliver Strewe **98**tl and tr Oliver Strewe; b Bill Bachman **99**t E and D Hosking/FLPA; c Austin J Brown/Aviation Picture Library; bl and br Oliver Strewe **100**t and bl David Parker/SPL/PL; br Paul Nevin/PL **101**tl SPL/PL; tr NASA/FLPA; bl and br Peter Jarver/Backgrounds **102**tl Scott Camazine/PR; tr Colin Monteath/OSF; c and b Warren B White, Physical Oceanography Research Division, Scripps Institute of Oceanography **103**t Luiz Claudio Marigo/BCL; cr Michael Salas/Image Bank, Sydney; bl Tony Bee/PL; br Tim Davis/PR **104**tl NASA/SPL/PL; tr and c Warren Faidley/IS; bl National Weather Service/NOAA **105**tl Dr Richard Legeckis/SPL/PL; c Gary Lewis/PL; bl Massachusetts Institute of Technology; br Howard Houben/NASA Center for Mars Exploration **106-107** (by Jan Griffier) Guildhall Art Gallery, Corporation of London/Bridgeman **108**tl Mehau Kulyk/SPL/PL; tr Breck P Kent/AA/ES; b Francois Gohier/A **109**t Ron Sanford/IS; c Breck P Kent/AA/ES **110**t AA&A; c George Holton/PR **111**tl Patti Murray/AA/ES; tr Dr Eckart Pott/BCL; b Kunsthistorisches Museum, Vienna/Bridgeman/PL **112**tl Ferrero/Labat/A; cr Alan Root/SAL/OSF **113**tl Breck P Kent/AA/ES; tr Zig Leszczynski/AA/ES; b SPL/PL **114**t Australian National Maritime Museum; bl GI Bernard/NHPA; br RJ Erwin/PR **115**t R Cechet/Antarctic Division; c D Parer and E Parer-Cook/A; b Neville Coleman's Underwater Geographic Pty Ltd/Australian Institute of Marine Sciences **116**tl Mark Marten/NASA/PR; cr Paul Nevin/PL; b Spencer Swanger/TS **117** Sepia Photo/ME **118** David Weintraub/PL **119** Baum and Henbest/SPL/PL **120**tr Gary Parker/SPL/PL; c Simon Fraser/SPL/PL; bl Carlos Goldin/SPL/PL **121**t Nick Gordon/SAL/OSF; b Jan Tove Johansson/PE **122**t Sinclair Stammers/SPL/PL; b Geoff Renner/Robert Harding Picture Library **123**t Hank Morgan/SPL/PL; b Robert M Carey, NOAA/SPL/PL **124**tl Y Lanceau/A; cl Carlos Goldin/SPL/PL; br Andy Price/BCL **125**cl Clive Collins; bl Hans-Peter Merten/BCL; br John Mead/SPL/PL **126**t Simon Fraser/SPL/PL; c Adenis/Sipa Press/PL; b UK Meteorological/SPL/PL **127**t SPL/PL; c Gunter Ziesler/BCL; b Adenis/Sipa Press/PL **128-129** Frank Grant/PL **130**t Edward Lettau/PR; b Jean-Loup Charmet **131**t Paul Thompson/PL; c ME; b Paul Trummer/Image Bank **132**tr Andy Price/BCL; c F Lechenet/Explorer/A **133**t Frank Grant/PL; c Aline Perier/A; b Christine Osborne Pictures **134**t ME; c Tony Stone Worldwide/PL; b Norbert Rosing/OSF **135**t Erwin and Peggy Bauer/BCL; b Frank Grant/IS **136**t George Band/Royal Geographical Society; b Sorrel Wilby/Wild Side **137**tl Edward Cross/PL; tr Tony Stone Worldwide/PL; b S Fiore/Explorer/A **138**t Ronald Sheridan/AA&A; bc Nino Marshall/PL; br Jean-Loup Charmet **139**t Everett Collection; c Hall of History Foundation, Schenectady NY; b courtesy Hydro-Electric Commission, Tasmania **140**t Jean-Loup Charmet/SPL/PL; c Kevin Schafer/TS; b Austin J Brown/Aviation Picture Library **141**t Jocelyn Burt/PL; b Dennis Sarson/LT **142-143** Graham Robertson **144**tl Renee Lynn/PR; tr Wendy Shattil and Bob Rozinski/TS; b A Wharton/FLPA **145**t Wayne Aldridge/IS; c Jiri Lochman/LT; b Nikita Ovsyanikov/PE **146**tl Erwin and Peggy Bauer/BCL; bl Daniel J Cox/PL; bc Dr Paul A Zahl/PR **148**tl Karl H Switak/PR; tr Suzanne L Collins and Joseph T Collins/PR; c Kenneth W Fink/PR **150**t Robert Tyrell/OSF/PL; bl Wayne Lawler/A; br John Lythgoe/PE **151** Michael Fogden/OSF **152**t Leo Meier; bl Robin Smith/PL; br M Stoklos/AA/ES **153** Esther Beaton/Terra Australis PhotoAgency **154** Claude Steelman/SAL/OSF **155**tl Jack Mackinnon/PE; tr Stephen Dalton/NHPA; b J Cancalosi/A **156**t John Downer/PE; b John Shaw/BCL **157** Philip Craven/Robert Harding Picture Library **158**t Frank Schneidermeyer/OSF; bl John Lythgoe/PE; br Richard Reid/AA/ES **159** Jane Burton/BCL **160**t Jean-Paul Ferrero/A; b Stephen Krasemann/NHPA **161**l Ron Austing/PR; r S Nielsen/BCL **162**t C Nuridsany&M Perennou/SPL/PL; b Michael Giannechini/PR **163** Erwin and Peggy Bauer/BCL **165**tl Norbert Schwirtz/BCL; tr John Shaw/A; c Daniel J Cox/OSF; b S Nielsen/BCL **166**tl Keith Gunnar/BCL; tr and c Hans Reinhard/BCL **167**l D and M Plage/SAL/OSF; r Keith Gunnar/BCL **168**tl Mark Newman/A; bl John Shaw/BCL; br Jeff Lepore/PR **169** David E Myers/Tony Stone Worldwide/PL **170**tl Dr Nick Gales/LT; tr Rick Price/SAL/OSF; c Tui De Roy/A **171** PV Tearle/PE **172**tl Bud Lehnhausen/PR; bl Rob Jung/PL; br Tom McHugh/PR **173** Deeble and Stone/SAL/OSF **174-175** Fred K Smith **178-179** Len Stewart/LT **179**ti John Gerlach/TS; bi Ron Oulds/BCL **180**t Hiroshi Higuchi/PL; c Anthony Bannister/NHPA; b John Shaw/NHPA **181**t Jan Tove Johansson/PE; b Christopher Wood Gallery/Bridgeman **182**t Michael J Howell/IS; b NRSC Ltd/SPL/PL **183**t Andre Baertschi/PE; b Mark Newman/FLPA **184**b Mike Coltman/PE **184-185** David E Rowley/PE **185**b Renee Lynn/PR **186**b ME **186-187** Eckart Pott/BCL **187**c L West/FLPA; b John Gerlach/TS **188-189** Peter Jarver/Backgrounds **189**ti Gordon Garradd/PL; bi Esther Beaton/Terra Australis Photo Agency **190**t Philip Quirk/W; b ME **191**t Ron Dorhan/PL; b Hermitage, St Petersburg/Bridgeman **192**b David Barrett/PE **192-193** Steve Nicholls/PE **194**t Jan Tove Johansson/PE; b ME **195**t GA Maclean/OSF; b Granger **196**t Trevor Worden/PL; b ME **197**t Len Stewart/LT; b GR Roberts/Documentary Photographs **198**t Gary D McMichael/PL; b B Cosgrove/FLPA **199**t Grahame McConnell/PL; b Peter Jarver/Backgrounds **200**b Austin J Brown/Aviation Picture Library **200-201** Peter Mackey **201**b Peter Jarver/W **202**b David Parker/SPL/PL **202-203** Kenneth D Langford/TS **203**b Bob Firth/IS **204**t Grahame McConnell/PL; b The MAAS Gallery, London/Bridgeman **205**t Paul Thompson/PL; b Granger **206**t Tom Keating/W; b David Miller **207**t Fritz Prenzel/BCL; b Jiri Lochman/LT **208**t Peter Scoones/PE; b Stan Osolinski/OSF **209**t B Cosgrove/FLPA; b Dennis Sarson/LT **210**b C Monteath/Hedgehog House, New Zealand/Explorer/A **210-211** Dr Robert Spicer/SPL/PL **211**b Doug Allan/OSF **212**t Greg Hard/W; b David Miller **213**t Allan G Potts/BCL; b John Lythgoe/PE **214** GR Roberts/Documentary Photographs **215**t Trevor Worden/PL; b D Parer and E Parer-Cook/A **216**t S Nielsen/BCL; b ME **217**t B Cosgrove/FLPA; b Dario Perla/IS **218-219** FPG International/Austral **219**ti Kurt Vollmer/PL; bi Bob Pool/TS **220**b Renee Lynn/PR **220-221** Raymond de Berquelle/PL **221**cl Fitzwilliam Museum/University of Cambridge/Bridgeman; br Dieter and Mary Plage/SAL/OSF **222**b Mary Clay/TS **222-223** John Shaw/TS **223**c Sharon Gerig/TS; b Jan Tove Johansson/PE **224**cr NCAR/TSADO/TS b Science Source/PR **224-225** Robert Stottlemyer/IS **225**c Spencer Swanger/TS **226**cl Anthony Bannister/NHPA; bl Science Photo Library/PL **226-227** Mary Clay/TS **227**c Mary Clay/TS; b ME **228**c Laurie Campbell/NHPA; b Fratelli Fabri, Milan/Bridgeman **228-229** Austin J Brown/Aviation Picture Library **229** UPI/Bettman/APL **230**t Doug Sokell/TS; c JM LaRoque/A; b Private Collection/Werner **231**t Bob Firth/IS; b JHC Wilson/Robert Harding Picture Library **232**b Peter Ward/BCL **232-233** Peter Jarver/Backgrounds **233**c Bill Bachman/PR **234**t Carol Hughes/BCL; b Nigel Dennis/NHPA **235**t Richard Packwood/OSF; b Bettman/APL **236-237** Warren Faidley/IS **237**ti Tony Bee/PL; bi Warren Faidley/IS **238**b Ray Ellis/PR **238-239**Peter Jarver/W **239**c AA&A **240-241** Wm L Wantland/TS **241** Wm L Wantland/TS **242** Keith Kent/SPL/PL **243** Peter Jarver/W **244**b Granger **244-245** Howie Bluestein/Science Source/PR **245**b Merrilee Thomas/TS **246** Dr Scott Norquay/TS **247**t Buff Corsi/TS; b Jonathan Scott/ PE **248** Peter Jarver/Backgrounds **249**t Robin Smith/PL; b ME **250**c Warren Faidley/IS; b NOAA/TS **250-251** NASA/SPL/PL **251**c Galen Rowell/Mountain Light/Explorer/A **252**t Robin Moyer/Black Star; b ME **253**t Byron Augustin/TS; b Flip Schulke/PE **254-255** Dr ER Degginger **255**ti Dr Scott Nielsen/BCL; bi Kenneth D Langford/TS **256**b RP Lawrence/FLPA **256-257** Richard R Hansen/PR **257**c Daniel J Cox/Natural Exposures; b Image Select **258**t David Miller; b Jerry Schad/Science Source/PR **259**t George Post/SPL/PL; b Gordon Garradd/PL **260**t John Shaw/TS; b ME **261** D Parer and E Parer-Cook/A **262**b SPL/PL **262-263** Silvestris/FLPA **263**b David Miller **264** Peka Parviainen/SPL/PL **265**t Arthur Gloor/AA/ES; b ME **266**b Chad Ehlers/IS **266-267** Ulf E Wallin/Stock Photos/Image Bank **267**c ME **268**b Clive Collins **268-269** Peter Solness/W **269**c NASA/SPL/PL **270**b Robert M Carey/NOAA/SPL/PL **270-271**David A Ponton/PE **271**c M and K Krafft/A **272-273** ME

ILLUSTRATION AND MAP CREDITS
Mike Lamble: reference banding for 218-235 and 254-271
Rob Mancini: 142-173
Ngaire Sales: 174-271, reference banding for 178-217 and 236-253
Genevieve Wallace: 272-288
Mark Watson: 38, 58, 59, 112, 118, all maps 142-173
Rod Westblade: 26, 32, 33, 37, 39, 47, 48-49, 50, 53, 54, 108-109, 110, 117, 119, 120, 123, 125
David Wood: 25, 29, 30, 34, 35, 45

JACKET: Front: The Science Museum/Science and Society Picture Library; Scala; CC Lockwood/BCL; Nigel J Dennis/NHPA
Back: Nigel J Dennis/NHPA; Mary Clay/TS; Johnny Johnson/BCL; Hillary Wilkes/IS; Scala; Erwin and Peggy Bauer/BCL; Stan Osolinski/OSF **Front flap:** Joe McDonald/TS; ME; Keith Gunnar/BCL **Back flap:** Granger